The ASQ Certified Six Sigma Yellow Belt Study Guide

The ASQ Certified Six Sigma Yellow Belt Study Guide

Second Edition

Erica L. Farmer
Grace L. Duffy

ASQExcellence
Milwaukee, Wisconsin

Published by ASQExcellence, Milwaukee, WI
Produced and distributed by Quality Press, ASQ, Milwaukee, WI
 Published 2017, 2022
Printed in the United States of America
26 25 24 23 22 SWY 5 4 3 2 1

Library of Congress Cataloging-in-Publication Data

Names: Farmer, Erica L., author. | Duffy, Grace L., author.
Title: The ASQ certified six sigma yellow belt study guide, second edition / Erica L. Farmer and Grace L. Duffy.
Description: Includes bibliographical references and index. | Milwaukee, WI: ASQExcellence, 2022.
Identifiers: LCCN: 2022943933 | ISBN: 978-1-63694-032-8 (paperback) | 978-1-63694-033-5 (pdf) | 978-1-63694-034-2 (epub)
Subjects: LCSH Six sigma (Quality control standard)—Examinations—Study guides. | Production management—Examinations—Study guides. | Quality control—Statistical methods—Examinations—Study guides. | BISAC BUSINESS & ECONOMICS / Quality Control | STUDY AIDS / Professional
Classification: LCC TS156.17.S59 F37 2022 | DDC 658.5/620218—dc23

ASQ and ASQExcellence advance individual, organizational, and community excellence worldwide through learning, quality improvement, and knowledge exchange.

Bookstores, wholesalers, schools, libraries, and organizations: Quality Press and ASQExcellence books are available at quantity discounts with bulk purchases for business, trade, or educational, uses. For more information, please contact Quality Press at 800-248-1946 or books@asq.org.

To place orders or to browse the full selection of Quality Press and ASQExcellence titles, visit our website at: http://www.asq.org/quality-press.

ASQ
Quality Press
600 N. Plankinton Ave.
Milwaukee, WI 53203-2914
Email: books@asq.org
Excellence Through Quality™

ASQE™
Inspire Excellence

As always, we thank our significant others, Mike Farmer and John Duffy, for their patience and support while we disappeared into the office to write test items and solutions to assist our peers in successfully preparing for the ASQ Six Sigma Yellow Belt certification examination.

Table of Contents

Preface

The *ASQ Certified Six Sigma Yellow Belt Study Guide* is designed to aid those taking the ASQ Certified Six Sigma Yellow Belt (CSSYB) examination and is best used in conjunction with *The Certified Six Sigma Yellow Belt Handbook* (Quality Press). Section 1 provides 161 practice questions organized according to the five parts of the Yellow Belt Body of Knowledge (BoK) effective from March 2022 until March 2027. Section 2 gives the reader 139 additional practice questions from each of the five parts, in a randomized order.

For every question in both sections, detailed solutions are provided that explain why each answer is the correct one and which section of the BoK the question corresponds to so that any further study needed can be focused on specific sections. The provided solutions are taken from the same references used by the ASQ Item Writing teams who create the test items used in the formal Six Sigma Yellow Belt certification examination. A list of these references is included prior to the first set of questions and available at http://asq.org/cert.

A secondary audience for this study guide is those taking exams for ASQ certifications whose BoKs have some crossover with the CSSYB. Namely, the Certified Six Sigma Green Belt (CSSGB), Certified Six Sigma Black Belt (CSSBB), Certified Quality Improvement Associate (CQIA), Certified Quality Process Analyst (CQPA), and Certified Manager of Quality/Organizational Excellence (CMQ/OE). Using this guide in studying for any of these exams is extremely useful, particularly for the quality tools and team portions of the BoKs.

Unlike other resources on the market, all these questions and solutions were developed specifically to address the 2022 Six Sigma Yellow Belt Body of Knowledge and help those studying for it, including considering the proper depth of knowledge and required levels of cognition. Questions and solutions from the 2015 edition of this study guide have been included where they still address the 2022 Body of Knowledge. New items were created to fill gaps in the new BoK.

As a reminder, practice and sample test questions, such as those in this study guide, cannot be taken into ASQ certification exam rooms. The exams are open book, however, so it is highly recommended that you do take the references listed in this study guide into the examination room with you to look up or verify any answers as you work the exam questions.

We welcome your feedback and suggestions for improvement. Please contact us at books@asq.org so we may incorporate any suggestion for improvement into future printings or editions of this study guide.

Erica L. Farmer
Grace L. Duffy

Acknowledgments

The authors acknowledge the contributions of the authors who prepared the reference texts used to create the ASQ Six Sigma Yellow Belt certification examination and from which content was drawn to provide solutions to the sample test items in this study guide. The reference texts and authors are listed in detail prior to the first set of questions.

Also to be acknowledged are the Quality Press editors and staff, who graciously worked with the authors to make this study guide available to our readers.

American Society for Quality (ASQ) 2022 CERTIFIED SIX SIGMA YELLOW BELT (SSYB) BODY OF KNOWLEDGE

The topics in this Body of Knowledge include additional detail in the form of subtext explanations and the cognitive level at which test questions will be written. This information will provide guidance for the candidate preparing to take the exam. The subtext is not intended to limit the subject matter or be all-inclusive of what might be covered in an exam. It is meant to clarify the type of content to be included in the exam. The descriptor in parentheses at the end of each entry refers to the maximum cognitive level at which the topic will be tested. A complete description of cognitive levels is provided at the end of this document.

I. Six Sigma Fundamentals (42 Questions)

A. Six sigma foundations and principles
Describe the purpose of six sigma (reducing variation), its methodology (DMAIC) and its evolution from quality. Describe the value of six sigma to the organization as a whole. (Understand)

B. Lean foundations and principles
Describe the purpose of lean (waste elimination) and its methodologies (just-in-time, poka-yoke, kanban, value-stream mapping). Describe the value of lean to the organization as a whole. (Understand)

C. Six sigma roles and responsibilities
Define and describe the roles and responsibilities of six sigma team members (i.e., individual team members, yellow belt, green belt, black belt, master black belt, process owner, champion, sponsor). (Understand)

D. Team basics

1. Types of teams
Identify the various types of teams that operate within an organization (i.e., continuous improvement, self-managed, and cross-functional) and their value. (Understand)

2. Stages of development
   Describe the various stages of team evolution: forming, storming, norming, performing, and adjourning. (Understand)

3. Decision-making tools
   Define and apply decision-making tools such as brainstorming, multivoting, and nominal group technique (NGT). (Apply)

4. Communication methods
   Explain how teams use agendas, meeting minutes, and project status reports, and how they support project success. (Apply)

E. Quality tools and six sigma metrics

1. Quality tools
   Select and use these tools throughout the DMAIC process: Pareto charts, cause and effect diagrams, flowcharts, run charts, check sheets, scatter diagram, and histograms. (Apply)

2. Six sigma metrics
   Select and use these metrics throughout the DMAIC process: defects per unit (DPU), defects per million opportunities (DPMO), rolled throughput yield (RTY), cycle time, and cost of poor quality (COPQ). (Apply)

II. Define Phase (26 Questions)

A. Project identification

1. Voice of the customer
   Define the voice of the customer and describe how customer needs are translated into quantifiable, critical-to-quality (CTQ) characteristics. (Understand)

2. Project selection
   Describe how projects are identified and selected as suitable for a six sigma project using the DMAIC methodology. (Understand)

3. Stakeholder analysis
   Identify end users, subject matter experts, process owners, and other people or factors that will be affected by a project and describe how each of them can influence the project. (Understand)

4. Process inputs and outputs
   Use SIPOC (suppliers, inputs, process, outputs, customers) to identify and define important elements of a process. (Apply)

5. Supply chain management
   Understand supply chain management and how it relates to project management. (Understand)

B. Project management (PM) basics

1. Project charter
   Describe the purpose of a charter and its components: problem statement, project scope, baseline data, and project goal. (Understand)
2. Communication plan
   Explain the purpose and benefits of a communication plan and how it can impact the success of the project. (Understand)
3. Project planning
   Define work breakdown structure (WBS) and Gantt charts and describe how they are used to plan and monitor projects. (Understand)
4. Project management tools
   Select and use various PM tools: activity network diagrams, affinity diagrams, matrix charts, relations charts, and tree diagrams. (Understand)
5. Phase reviews
   Explain how tollgate or phase reviews are used throughout the DMAIC lifecycle. (Understand)

III. Measure Phase (30 Questions)

A. Basic statistics

Define, calculate, and interpret measures of central tendency (mean, median, mode) and understand measures of dispersion (standard deviation, range, variance). (Apply)

B. Data collection

1. Data collection plans
   Describe the critical elements of a data collection plan, including an operational definition, data sources, the method to be used for gathering data, and how frequently it will be gathered. Describe why data collection plans are important. (Understand)
2. Qualitative and quantitative data
   Define and distinguish between these types of data. (Understand)
3. Data collection techniques
   Use various data collection techniques, including surveys, interviews, check sheets, and checklists to gather data that contribute to the process being improved. (Apply)

C. Measurement system analysis (MSA)

1. MSA terms
   Define precision, accuracy, bias, linearity, and stability and describe how these terms are applied in the measurement phase. (Understand)

2. Gage repeatability & reproducibility (GR&R)
   Describe and distinguish between repeatability and reproducibility and describe how and why GR&R is used in the measurement phase. (Understand)

IV. Analyze Phase (37 Questions)

A. Process analysis tools

1. Lean tools
   Define how 5S and value analysis can be used to identify and eliminate waste. (Understand)

2. Failure mode and effect analysis (FMEA)
   Relate the elements of severity, occurrence, and detection and determine how they are used to calculate the risk priority number. Demonstrate how FMEA can be used to identify potential failures in a process. (Apply)

B. Root cause analysis

Describe how the 5 Whys, process mapping, 8D, force-field analysis and matrix charts can be used to identify the root causes of a problem. (Understand)

C. Corrective action

Explain and apply elements of the corrective action process: identify the problem, contain the problem (interim action), determine the causes of the problem, propose solutions to eliminate it or prevent its recurrence (permanent action), verify that the solutions are implemented, and confirm their effectiveness (validation). (Apply)

D. Preventive action

Explain and apply elements of a preventive action process: understand various process analysis techniques to identify potential failures, defects, or process deficiencies; improve the process (e.g., understand error- or mistake-proofing devices or methods, initiate procedural changes), and verify the effectiveness of the preventive action. (Apply)

E. Data analysis

1. Basic distribution types
   Define and distinguish between normal and binomial distributions and describe how their shapes (skewed and bimodal) can affect data interpretation. (Understand)

2. Common and special cause variation
   Describe and distinguish between these types of variation. (Understand)

F. Correlation and regression

1. Correlation
   Describe how correlation is used to identify relationships between variables. (Understand)

2. Regression
   Describe how regression analysis is used to predict outcomes. (Understand)

G. Hypothesis testing
Define and distinguish between hypothesis terms (i.e., null and alternative, type I and type II error, p-value and power). (Understand)

V. Improve and Control Phases (26 Questions)

A. Improvement techniques

1. Kaizen and kaizen blitz
   Define and distinguish between these two methods and describe how they can be used to make improvements to any process in an organization. (Understand)

2. Plan-do-check-act (PDCA) cycle
   Define and distinguish between the steps in this process improvement tool. (Understand)

3. Cost-benefit analysis
   Explain the importance of this analysis and how it is used in the improve phase. (Understand)

B. Control tools and documentation

1. Control plan
   Describe the importance of a control plan for maintaining improvements. (Understand)

2. Control charts
   Describe how $\overline{X}$ charts are used for monitoring and sustaining improved processes. (Understand)

3. Document control
   Describe the importance of documenting changes to a process and communicating those changes to stakeholders. (Understand)
4. Work instructions and standard operating procedures (SOPs)
   Understand the purpose and use of work instructions and SOPs. (Understand)

LEVELS OF COGNITION

Based on Bloom's Taxonomy—Revised (2001)

In addition to **content** specifics, the subtext for each topic in this BoK also indicates the intended **complexity level** of the test questions for that topic. These levels are based on "Levels of Cognition" (from Bloom's Taxonomy—Revised, 2001) and are presented below in rank order, from least complex to most complex.

Remember

Recall or recognize terms, definitions, facts, ideas, materials, patterns, sequences, methods, principles, etc.

Understand

Read and understand descriptions, communications, reports, tables, diagrams, directions, regulations, etc.

Apply

Know when and how to use ideas, procedures, methods, formulas, principles, theories, etc.

Analyze

Break down information into its constituent parts and recognize their relationship to one another and how they are organized; identify sublevel factors or salient data from a complex scenario.

Evaluate

Make judgments about the value of proposed ideas, solutions, etc., by comparing the proposal to specific criteria or standards.

Create

Put parts or elements together in such a way as to reveal a pattern or structure not clearly there before; identify which data or information from a complex set is appropriate to examine further or from which supported conclusions can be drawn.

ASQ Six Sigma Yellow Belt Certification Preparation Recommended References

1. Duffy, Grace L. and Sandra L. Furterer. *The Certified Quality Improvement Associate Handbook,* 4th ed. Milwaukee: ASQExcellence, 2020.
2. Munro, Roderick A., Govindarajan Ramu, and Daniel J. Zrymiak. *The Certified Six Sigma Green Belt Handbook,* 2nd ed. Milwaukee: ASQ Quality Press, 2015.
3. Ramu, Govind. *The Certified Six Sigma Yellow Belt Handbook,* 2nd ed. Milwaukee: ASQ Quality Press, 2022.
4. Munro, Roderick A. *Six Sigma for the Shop Floor: A Pocket Guide.* Milwaukee: ASQ Quality Press, 2002.
5. Scholtes, Peter R., Brian L. Joiner, and Barbara J. Streibel. *The Team Handbook,* 3rd ed. Oriel, 2003.
6. Tague, Nancy R. *The Quality Toolbox,* 2nd ed. Milwaukee: ASQ Quality Press, 2005.

Section 1
Sample Questions by BoK

Part I	Six Sigma Fundamentals
Part II	Define Phase
Part III	Measure Phase
Part IV	Analyze Phase
Part V	Improve and Control Phases

Section 1 is divided into five parts, one for each section in the ASQ Certified Six Sigma Yellow Belt Body of Knowledge. In each part there is a set of questions followed by detailed solutions, including reference (see p. xxi) and section from which the solution is taken.

Part I

Six Sigma Fundamentals

(42 questions)

QUESTIONS

1. The purpose of Six Sigma is to:
   a. eliminate waste.
   b. use control charts.
   c. count defects.
   d. reduce variation.
2. A significant value of Six Sigma is to:
   a. continually improve all aspects of the organization.
   b. provide issues for project teams to resolve.
   c. illustrate the sigma value of defect levels.
   d. replace the plan-do-check-act cycle.
3. The DMAIC model stands for:
   a. design, measure, analyze, improve, and control.
   b. define, monitor, assess, implement, and control.
   c. define, measure, analyze, improve, and control.
   d. design, monitor, assess, implement, and contain.
4. ____________ developed quality control charts in the 1920s to identify special causes affecting processes and their outcomes.
   a. Ishikawa
   b. Feigenbaum
   c. Deming
   d. Shewhart

5. The quality concept of Six Sigma was initiated by which company in the 1980s?

   a. International Business Machines

   b. General Electric

   c. AT&T

   d. Motorola

6. The purpose of lean is to ____________ in the value stream and to ____________ our customers.

   a. fix errors, sell services

   b. position inventory, supply products and services

   c. reduce waste, provide maximum value

   d. create transparency, enhance market share

7. A team wishing to mistake-proof the assembly of a new piece of equipment might choose which of the following?

   a. Kaizen

   b. Check sheet

   c. Gantt chart

   d. Poka-yoke

8. Which of the following is "a systematic approach in identifying and eliminating waste (non-value-added activities) through continuous improvement by flowing the product at the pull of the customer in pursuit of perfection"?

   a. PDSA

   b. Gemba

   c. Lean

   d. Six Sigma

9. Which of the following describes "a production strategy promoted by Toyota, and now applied to many organizations, that strives to improve business return on investment by reducing in-process inventory and associated carrying costs"?

   a. Just-in-time

   b. Heijunka

   c. Quality circles

   d. Return on investment

10. To whom should the team leader speak to free up financial resources for a required software upgrade?

    a. Team facilitator

    b. Project sponsor

    c. Department manager

    d. Purchasing director

11. Who is the individual trained to have awareness in Six Sigma methodologies and understanding of basic statistical tools and process improvement techniques?

    a. Yellow Belt

    b. Sponsor

    c. Leader

    d. Black Belt

12. Of the following, who should select team members and ensure that improvements are sustained?

    a. Black Belt

    b. Facilitator

    c. Process owner

    d. Sponsor

13. Which level of Six Sigma Belts is responsible for facilitating Six Sigma methodologies, statistical tools, basic financial tools, change management, risk assessment, and project management activities?

    a. White

    b. Yellow

    c. Green

    d. Black

14. Company ABC has just partnered with company HAL to provide a new online information service. Which of the following would be the best team structure to assure success?

    a. Process improvement

    b. Natural

c. Project

d. Virtual

15. Which of the following types of team structures involves a broader spectrum of process responsibility and ownership?

a. Process improvement

b. Self-directed

c. Quality circles

d. Work cell

16. A ____________ team enables people from all over the globe to meet via teleconference and internet tools.

a. natural

b. corrective action

c. virtual

d. work group

17. Which of the following is a common action exhibited during the norming stage of team growth?

a. Diagnosing and solving problems

b. Becoming impatient with the lack of progress

c. Testing the leader's guidance

d. Reconciling competing loyalties and responsibilities

18. One of the things a team leader should do in a team's first meeting is:

a. set project deadlines and reporting.

b. train team members on quality tools.

c. flowchart the organizational process.

d. define roles and responsibilities.

19. Frustration and resistance to the assigned task are a symptom of the ____________ stage of team growth.

a. storming

b. performing

c. forming

d. norming

20. During which stage of team development do members reconcile competing loyalties and responsibilities?

    a. Adjourning

    b. Storming

    c. Norming

    d. Forming

21. Given that a list of seven acceptable software tools has been researched for a new project implementation, which of the following tools should the team use to select the most desirable choice?

    a. Cause-and-effect diagram

    b. Pareto chart

    c. Nominal group technique

    d. Brainstorming

22. The facilitator of a brainstorming session should do which of the following?

    a. Assign content responsibilities to members

    b. Select the best ideas to be reported to management

    c. Solicit involvement from group members

    d. Critique new ideas presented during the session

23. Which of the following is a reason why multivoting is preferable to straight voting when making a decision?

    a. It can be performed by an individual using a decision matrix rather than involving a whole team.

    b. It allows an item that is favored by all, but not the top choice of any, to rise to the top.

    c. Multivoting must be performed over a sequence of votes, whereas straight voting is performed only once.

    d. Hidden agendas among team members are less likely to influence the results using multivoting.

24. When participants might feel safer contributing ideas anonymously, which of the following decision-making tools would a facilitator choose?

    a. Root cause analysis

    b. Brainwriting

    c. Nominal group technique

    d. SWOT

25. Which of the following items should be included in a meeting agenda?

    a. Process flowchart

    b. Discussion topics

    c. Project solutions

    d. Team ground rules

26. Any unacceptable project variances and risks should be addressed early in the project by reviewing the ____________ and taking appropriate actions.

    a. project charter

    b. senior management briefing

    c. project status report

    d. voice of the customer

27. Which of the following is an effective way to ensure that key decisions made and actions agreed to in team meetings are formally recorded?

    a. Policies

    b. Standards

    c. Minutes

    d. Ground rules

28. When should the team review their project status and communicate with major stakeholders?

    a. At the end of each major step in their project or work plans

    b. Before each team meeting to ensure they are on track

    c. After each team meeting to verify accuracy of any decisions made

    d. Only at the end of the project to share results and lessons learned

29. Flowcharts are most effectively developed by:
    a. outside consultants without company bias.
    b. upper management responsible for the process.
    c. industrial engineers with training in graphics.
    d. people involved in the process being studied.

30. Which of the following tools is most useful for viewing the frequency or number of observations of a specific set of data?
    a. Cause-and-effect diagram
    b. Histogram
    c. Fishbone diagram
    d. Scatter diagram

31. A cause-and-effect diagram is also known as a:
    a. histogram.
    b. run chart.
    c. Gantt chart.
    d. fishbone diagram.

32. Which of the following tools would be used to study observed data for trends or patterns over a specified period of time?
    a. Scatter diagram
    b. Run chart
    c. Affinity diagram
    d. Histogram

33. Which of the following basic quality tools is used to display what happens to one variable when another variable is changed?
    a. Cause-and-effect diagram
    b. Histogram
    c. Scatter diagram
    d. Flowchart

34. When collecting data on the frequency or patterns of events, problems, defects, defect location, or defect causes, which of the following tools should be employed?

    a. Histogram

    b. Run chart

    c. L-shaped matrix

    d. Check sheet

35. When first learning to use the cause-and-effect diagram, many teams will use which of the following approaches to label the major "fishbones," or categories, on the diagram?

    a. 6 M's

    b. 5 Whys

    c. 8D

    d. 7 tools

36. Run charts differ from control charts because they:

    a. use the *y*-axis to measure time sequence.

    b. do not have statistical control limits to monitor variation.

    c. use specification limits rather than control limits.

    d. are used for larger populations of data.

Questions 37, 38, and 39 are based on the following information.

A medical billing department conducts a monthly audit of invoices to assess accuracy in billing practices. The areas being assessed for accuracy are signature, patient last name, telephone number, and amount billed. An auditor reviewed 100 invoices and found the following: five incorrect signatures, three missing last names, one missing telephone number, and three incorrect amounts billed.

37. Which of the following is correct if measuring defects per unit?

    a. 0.12

    b. 0.48

    c. 0.04

    d. 0.16

38. Which of the following is correct if measuring defects per million opportunities?

    a. 120,000

    b. 40,000

c. 30,000

d. 160,000

39. The sigma level for the medical billing process is

a. ~3.375.

b. ~3.250.

c. ~2.500.

d. ~2.700.

40. A mortgage lending company wanted to know the percentage of applications that made it through the entire process defect free. Which of the following is used to calculate this metric?

a. Rolled process yield

b. Rolled throughput yield

c. Rolled through yield

d. Rolled yield

41. Using the following information, what is the rolled throughput yield for the mortgage application process?

Sales process yield = 95%

Processing process yield = 89%

Underwriting process yield = 96%

Closing process yield = 93%

a. 75.5%

b. 95.5%

c. 94.5%

d. 75.0%

42. Which of the following concepts describes costs incurred because things are not done right the first time and every time?

a. Project outcome cost analysis

b. Taguchi method

c. Cost of poor quality

d. Critical-to-quality analysis

SOLUTIONS

1. d; The purpose of Six Sigma is reducing variation. While variability is inherent, organizations strive to keep it as low as possible in order to deliver products and services that meet or exceed customer expectations. Highly variable products and services result in poor quality, and the organization must spend money to correct the defect. [I.A] (Reference #3, I.—A)

2. a; The real power of Six Sigma is the use of many parts or elements of other methods that have been proven to work, in tandem with managerial focus, to create an organizational network of activities that support the efforts to continually improve all aspects of the organization, in conjunction with standard accounting practices that demonstrate the impact of continual improvement and variation reduction on the organization's bottom-line. [I.A] (Reference #2, Ch.1—A.1)

3. c; The DMAIC model stands for define, measure, analyze, improve, and control. [I.A] (Reference #4, The DMAIC Model)

4. d; In the mid-1920s a young engineer named Walter Shewhart devised a technique of using graphs to monitor a process in order to identify whether that process was acting in a predictable manner or if what he termed *special causes* were affecting the process. These charts became known as *quality control charts* (the *p* chart was the first to be used); however, today we sometimes call them *process behavior charts*, as we want to look at what the process is doing in relation to statistical probabilities. Many other tools and techniques have been developed since then, known by a long list of names. [I.A] (Reference #2, Ch.1—A.1)

5. d; After Motorola started promoting their Six Sigma methodology in the late 1980s, there have been many skeptical of its true value. Even Jack Welch of General Electric (GE) initially dismissed the idea of Six Sigma as a passing fad in the early 1990s. However, once GE had a successful launch in one of its divisions, Six Sigma quickly became a driving force in the mid to late 1990s that started spreading across various industries. [I.A] (Reference #2, Ch.1—A.1)

6. c; The purpose of lean is to reduce waste in the value stream and provide maximum value to our customers. [I.B] (Reference #3, I.B)

7. d; *Poka-yoke*, a Japanese term for mistake-proofing or error-proofing, is a method used to prevent errors. [I.B] (Reference #3, I.B)

8. c; A definition of lean, used by the Manufacturing Extension Partnership (of NIST/MEP, part of the U.S. Department of Commerce), is "a systematic approach in identifying and eliminating waste (non-value-added activities) through continuous improvement by flowing the product at the pull of the customer

in pursuit of perfection." Lean focuses on value-added expenditure of resources from the customer's viewpoint. [I.B] (Reference #1, Ch.7—Lean Concepts and Tools)

9. a; Just-in-time (JIT), a production strategy promoted by Toyota, and now applied to many organizations, that strives to improve business return on investment by reducing in-process inventory and associated carrying costs. [I.B] (Reference #2, Ch.2—B)

10. b; The project sponsor:
    - Believes in the concept/idea
    - Has sound business acumen
    - Is willing to take risk and responsibility for outcomes
    - Has authority to approve needed resources
    - Will be listened to by upper management

    [I.C] (Reference #1, Ch.5—Team Roles and Responsibilities)

11. a; Individuals trained to have awareness in Six Sigma methodologies and understanding of basic statistical tools and process improvement techniques. This is not a full-time position. Organizations make this part of an existing job responsibility. Management and champion are also trained in Yellow Belt methodologies. [I.C] (Reference #3, I.C)

12. c; An individual responsible and accountable for the execution and results of a given process:
    - Selecting team members
    - Allocating resources for projects
    - Providing process knowledge
    - Reviewing process changes
    - Approving changes/supporting change management
    - Implementing change
    - Ensuring that improvements are sustained

    [I.C] (Reference #3, I.C t)

13. d; Black Belt individuals trained in Six Sigma methodologies, statistical tools, basic financial tools, change management, risk assessment, and project management, and well experienced in managing Black Belt projects. [I.C] (Reference #3, I.C t)

14. c; A project team is formed to achieve a specific mission. The project team's objective may be to create something new, such as a facility, product, or service. [I.D.1] (Reference #1, Ch.4—Types of Teams)

15. b; Self-directed (self-managed) teams are groups of employees authorized to make a wide range of decisions about how they will handle issues. Often called *high-performance work teams,* these teams offer employees a broader spectrum of responsibility and ownership of a process. [I.D.1] (Reference #1, Ch.4—Types of Teams)

16. c; Virtual teams enable people from all over the globe to meet via teleconferences, videoconferences, and internet tools such as shared computers. There are many benefits to virtual teaming, the most prevalent being reduced costs and real-time data sharing and updating. [I.D.1] (Reference #2, Ch.9—F)

17. d; During the *norming* stage, members reconcile competing loyalties and responsibilities. They accept the team, the team ground rules or "norms," their roles in the team, and the individuality of fellow members. [I.D.2] (Reference #5, p. 6–7)

18. d; Define roles. Discuss how the team will operate. Discuss the responsibilities of team members in contributing to understanding the problem or process and carrying out some of the data collection and analysis. [I.D.2] (Reference #5, p. 3–52)

19. a; *Storming* is probably the most difficult stage for the team. Storming includes these feelings—frustration and resistance to tasks and methods of work different from what each individual member is comfortable using. [I.D.2] (Reference #5, p. 6–5)

20. c; Stage 3: norming. During this stage, members reconcile competing loyalties and responsibilities. They accept the team, the team ground rules or "norms," their roles in the team, and the individuality of fellow members. [I.D.2] (Reference #5, p. 6–7)

21. c; The nominal group technique (NGT) is a structured process that identifies and ranks major problems or issues that need addressing. It can be used to identify the major strengths of a department or to make decisions by consensus when selecting problem solutions in a business. [I.D.3] (Reference #1, Ch.6—Team Decision-Making)

22. c; Invite people to call out their ideas. Record all ideas in words as close as possible to those used by the contributor. No discussion or evaluation of any kind is permitted. [I.D.3] (Reference #6, Ch.5—The Tools—Brainstorming)

23. b; Multivoting narrows a large list of possibilities down to a smaller list of the top priorities or to a final selection. Multivoting is preferable to straight voting because it allows an item that is favored by all, but not the top choice of any, to rise to the top. [I.D.3] (Reference #6, Ch.5—The Tools—Multivoting)

24. b; Brainwriting is a nonverbal form of brainstorming. Team members write down their ideas individually. Ideas are shared through an exchange of papers, and then additional ideas are written down.

    When to use: Try brainwriting instead of brainstorming for any of the following reasons:

    - When a topic is too controversial or emotionally charged for a verbal brainstorming session
    - When participants might feel safer contributing ideas anonymously
    - To encourage equal participation, typically when verbal brainstorming sessions are dominated by a few members
    - When some group members think better in silence
    - When ideas are likely to be complex and require detailed explanation

    [I.D.3] (Reference #6, Ch.5—The Tools—Brainwriting)

25. b; Agendas should include the following information:

    - Purpose of the meeting
    - Topics, including a sentence or two that defines each item and why it is being discussed
    - The lead person for each topic
    - Time estimates

    [I.D.4] (Reference #5, p. 3–2)

26. c; *Project status report.* This is a periodic status report created by the project manager or team leader and circulated to all team members and management that lists the status of the project, upcoming milestones, risks, and mitigation plans. Any unacceptable project variances and risks are addressed early in the project by reviewing the status reports and taking appropriate actions. [I.D.4] (Reference #3, I.D.4)

27. c; *Minutes.* The meeting minutes are a record of the meeting. Minutes are essential to ensure that key decisions made and the actions agreed on by the team members are formally recorded, and to keep the team members accountable. The minutes should be well drafted and unambiguous, and they

should indicate the date and time of the completed meeting, meeting host, attendees, topics covered, decisions made, actions assigned (what, who, when), minutes reviewed and approved, and scribe name. One might also include "parking lot" items that are pending detailed discussions. Meeting minutes are important to project continuity. [I.D.4] (Reference #3, I.D.4)

28. a; Take time periodically throughout the project to stop and ask, "How are we doing?" This can be one of the most important and difficult activities a team can undertake. Self-critique can help a team identify problems early before they become crises. We recommend that teams pause to review their work at the end of each major step in their work plans. [I.D.4] (Reference #5, p. 3–60)

29. d; Identify and involve in the flowcharting process all key people who are involved with the process. This includes those who do the work in the process: suppliers, customers, and supervisors. Involve them in the actual flowcharting sessions by interviewing them before the sessions and/or by showing them the developing flowchart between work sessions and obtaining their feedback. [I.E.1] (Reference #6, Ch.5—The Tools—Flowchart)

30. b; A *histogram* is a graphic representation (bar chart) used to plot the frequency with which different values of a given variable occur. Histograms are used to examine existing patterns, identify the range of variables, and suggest a central tendency in variables. [I.E.1] (Reference #1, Ch.9—Improvement Tools)

31. d; The *cause-and-effect diagram* graphically illustrates the relationship between a given outcome and all the factors that influence the outcome. It is sometimes called the *Ishikawa diagram* or *fishbone diagram*. [I.E.1] (Reference #1, Ch.10—Root Cause Analysis)

32. b; The *run chart* reveals patterns in the data over time. It is used when monitoring a continuous variable over time and when looking for patterns, such as cycles, trends, or changes in the average. [I.E.1] (Reference #6, Ch.5—The Tools—Run Chart)

33. c; A *scatter diagram* is a chart in which one variable is plotted against another to determine whether there is a correlation between the two variables. These diagrams are used to plot the distribution of information in two dimensions. Scatter diagrams are useful in rapidly screening for a relationship between two variables.

    A scatter diagram shows the pattern of relationships between two variables that are thought to be related. For example, is there a relationship between outside temperature and cases of the common cold? As temperatures drop, do colds increase? The more closely the points hug a diagonal line, the more likely it is that there is a one-to-one relationship.

The purpose of the scatter diagram is to display what happens to one variable when another variable is changed. The diagram is used to test a theory that the two variables are related. The slope of the diagram indicates the type of relationship that exists. [I.E.1] (Reference #1, Ch.9—Improvement Tools)

34. d; A *check sheet* is a structured, prepared form for collecting and analyzing data. This is a generic tool that can be adapted for a wide variety of purposes.

    When to use:

    - When data can be observed and collected repeatedly by the same person or at the same location
    - When collecting data on the frequency or patterns of events, problems, defects, defect location, defect causes, and so forth
    - When collecting data from a production process

    [I.E.1] (Reference #6, Ch.5—The Tools—Check Sheet)

35. a; When using the cause-and-effect diagram, it is best to try to keep an open mind and to work as a team to view and discuss what the system or process is doing. You want to capture everything you can about the process, looking for the real state of the system, not just what you think is happening. Besides using the five W's and one H in creating the cause-and-effect diagram, many people start with the six M's:

    1. Man (people—operator)
    2. Machine (equipment)
    3. Methods (operating procedures)
    4. Materials
    5. Measurement
    6. Mother nature (environment)

    [I.E.1] (Reference #4, Tools—Cause and Effect Diagram)

36. b; The *run chart* is used to identify patterns in process data. All of the individual observations are plotted in a time sequence, and a horizontal reference line is drawn at the median. A run chart is typically used when the subgroup size is one. When the subgroup size is greater than one, the subgroup means or medians are calculated and connected with a line, similar to a control chart. However, run charts are different from control charts (for example, $\bar{X}$ and R charts); *run charts do not have statistical control limits to monitor variation.* There are also related statistical tests that can be performed to detect any nonrandom behavior. [I.E.1] (Reference #3, Ch.5—E.1)

37. a; *Defects per unit* (DPU) is calculated as the total number of defects divided by the total number of products produced in some time period (for example, per day). [I.E.2] (Reference #3, I.E.2)

$$DPU = \frac{\text{Total number of defects found in a sample}}{\text{Sample size}}$$

$$DPU = \frac{12}{100} = .12$$

Sample size = 100 invoices

Total number of defects found in the sample = 5 + 3 + 1 + 3 = 12:

- 5 incorrect signatures
- 3 missing last names
- 1 missing telephone number
- 3 incorrect amounts billed

38. c; *Defects per million opportunities* (DPMO) is calculated as follows: DPMO = (Total number of defects divided by the total number of units, multiplied by the number of opportunities per unit) multiplied by 10^6 [I.E.2] (Reference #3, I.E.2)

$$DPMO = \frac{\text{Total number of defects found in a sample}}{\text{Total number of defect opportunities found in a sample}}$$

$$DPMO = \frac{12}{400} \times 1{,}000{,}000 = 30{,}000$$

Sample size = 100 invoices

Total number of defect opportunities found in a sample = 100 × 4 = 400

Total number of defects found in the sample = 12

39. a; A *sigma level* is 3.375. This represents the quality level of the process—the higher the sigma level, the higher the quality level and the fewer defectives per million units produced or service transactions rendered. There are two types of sigma level calculations: sigma level with Motorola 1.5 sigma shift and without the shift (most sigma conversion tables include the 1.5 sigma shift). To determine the sigma level, look up 30,000 DPMO in a Six Sigma conversion table. Find the closest DPMO and select the sigma level. In this case, the sigma level with a 1.5 sigma shift is ~3.375. [I.E.2] (Reference #3, I.E.2)

40. b; *Rolled throughput yield* (RTY) applies to the yield from a series of processes and is found by multiplying the individual process yields. [I.E.2] (Reference #3, I.E.2)

41. a; The *rolled throughput yield* (RTY) for the application process is 75.5%. RTY applies to the yield from a series of processes and is found by multiplying the individual process yields. The calculation used to determine RTY for the application process is as follows: $.96 \times .95 \times .93 \times .89 = .755$ or 75.5%. [I.E.2] (Reference #2, Ch.8—E.1)

42. c; A *cost-of-poor-quality analysis* is a way of studying a process's flowchart to identify potential problems. *Cost of poor quality* means costs incurred because things are not done right the first time and every time. The analysis helps a team look critically at individual steps of a process to find opportunities for improvement.

    When to use:

    - When flowcharting a process, to be sure that cost-of-poor-quality activities are included
    - After flowcharting a process, to identify problems, potential causes, and areas in which to concentrate improvement efforts

    [I.E.1] (Reference #6, The Tools—Cost of Poor Quality)

Part II

Define Phase

(26 questions)

QUESTIONS

1. Voice of the customer data gathering might include which of the following?

   a. Expired product reports

   b. Complaint logs

   c. Audit review

   d. Internal failure records

2. Disney World management asking attendees how Disney World could improve their visiting experience would be an example of:

   a. complaint management.

   b. supplier selection.

   c. customer research.

   d. process design.

3. Methods used to identify customers include which of the following?

   a. Requirements and measures tree

   b. Value stream mapping

   c. Tracking a product to delivery

   d. Nominal group technique

4. A sponsor would assign an improvement activity to a Six Sigma project team if the problem involved which of the following characteristics?

   a. Intense data analysis

   b. Focused waste reduction

c. Prevention cost identification

d. Value stream mapping

5. A more experienced cross-functional Six Sigma team will likely be assigned a project with a high level of ____________.

a. homogenous data

b. technical complexity

c. process stability

d. organizational agreement

6. When selecting a Six Sigma project, which of the following is the most critical characteristic?

a. Saving the most bottom-line dollars

b. Reducing waste and excess inventory

c. Receiving the highest value in multivoting

d. Getting the highest rating on an employee survey

7. A company that strives to reduce toxic emissions from its vehicles is directly addressing the needs of which stakeholder customer group?

a. Suppliers

b. Creditors

c. Community

d. Shareholders

8. Which of the following stakeholders would be most directly impacted by a reduction of internal failures in a rental car call center?

a. Stockholders

b. Employees

c. Consumers

d. Suppliers

9. If the Six Sigma improvement team wishes to gather direct feedback from someone who uses their product in their own home, which of the following would they interview?

a. End user

b. Distributor

c. Wholesaler

d. Process owner

10. When creating a SIPOC for a process under study, inputs that are identified come from ________________.

a. customers

b. employees

c. stakeholders

d. suppliers

11. Which of the following is used when it is not clear what the outputs of a process are?

a. Histogram

b. 5 Whys

c. SIPOC

d. DOE

12. A Six Sigma Yellow Belt team member has been asked to develop a high-level description of a process, including the inputs and outputs. Which of the following tools should they employ?

a. Affinity diagram

b. PERT

c. Run chart

d. SIPOC

13. Which of the following statements best describes supply change management?

a. The process of effectively monitoring an efficient delivery system

b. The process of effectively integrating and managing components of the supply chain

c. The process of effectively building collaborative partnerships with key suppliers and business partners

d. The process of effectively managing activities and events throughout the supply chain

14. There are a number of similarities between project management and supply chain management. Which of the following would apply primarily to supply chain management?

    a. Planning and scheduling

    b. Performance measurement

    c. Supplier integration

    d. Managing costs

15. Which of the following is a component of the project charter?

    a. Project scope

    b. Project escalation

    c. Project scheduling

    d. Project planning

16. A problem statement is defined as:

    a. the limitations of the project in terms of time, cost, and resources.

    b. the criteria and metrics that determine success.

    c. the summation of what needs to be improved.

    d. the when and where based on the customer's perspective.

17. An effective communication plan contains a communication protocol that addresses the following items:

    a. what, who, where, when, and how frequently.

    b. what, who, where, when, and work breakdown structure.

    c. what, who, where, when, and graphs.

    d. what, who, where, when, and capability planning.

18. Which of the following describes an update protocol (what, who, where, when, how, and how frequently), escalation procedure, escalation threshold (when to escalate), and feedback effectiveness verification for a Six Sigma process team?

    a. DMAIC process

    b. Communication plan

    c. Interrelationship digraph

    d. Critical-to-quality exercise

19. A Six Sigma project team has been formed and the team leader is conducting a brainstorming effort to develop a detailed project plan. Which of the following is a likely output of this brainstorming session?

    a. Gantt chart

    b. Work breakdown structure

    c. Communication plan

    d. Matrix chart

20. A graphical method for monitoring key activities and their duration is called a(n):

    a. matrix chart.

    b. activity network diagram.

    c. Gantt chart.

    d. affinity diagram.

21. Effective project planning requires several qualitative, or behavioral, skills. Which of the following is one of those skills?

    a. Communication

    b. Statistical process control

    c. Confrontation

    d. Siloed thinking

22. Which of the following is a technique for gathering and organizing large numbers of ideas or facts?

    a. Scatter plot

    b. Comment card

    c. Focus diagram

    d. Affinity diagram

23. The main benefit of creating an activity network diagram is the calculation of the:

    a. slack time.

    b. earliest finish.

    c. latest finish.

    d. critical path.

24. Which of the following tools is used to discover and illustrate relationships between two groups of items?

    a. Matrix diagram

    b. Force-field analysis

    c. Scatter diagram

    d. Is/is not matrix

25. Which of the following is an appropriate indication that it is time to conduct a phase review?

    a. When there is a change of management

    b. When significant conflicts occur among team members

    c. When the team has identified performance gaps and narrowed the focus of the project

    d. When the current budget cycle is completed

26. A Six Sigma project team has reached a point during the *define* phase where they must seek sponsor approval for additional resources. Which of the following should they request from the sponsor?

    a. Root cause analysis

    b. Activity network diagram

    c. Milestone review

    d. Project audit

SOLUTIONS

1. b; The voice of the customer (VOC) might include gathering and integrating survey data, focus group findings, lost customer analysis, lost bids analysis (potential customers), warranty data, complaint logs and field reports, and any other data and information that affect the customer's purchasing and relationship decisions. [II.A.1] (Reference #3, Ch.6—A.1)

2. c; The *voice of the customer* (VOC) is a process for capturing customer-related information. This process is proactive and continuously innovative to capture stated, unstated, and anticipated customer requirements, needs, and desires. The goal is to achieve customer loyalty and to build customer relationships. The VOC might include gathering and integrating survey data, focus group findings, lost customer analysis, lost bids analysis (potential customers), warranty data, complaint logs and field reports, and any other data and information that affect the customer's purchasing and relationship decisions. These days, with the surge in social media, organizations should also be capturing and analyzing data from social networks for the VOC. [II.A.1] (Reference #3, II.A.1)

3. c; Methods used to identify customers include:

    - Brainstorming
    - SIPOC
    - Marketing analysis data
    - Tracking a product or service to delivery

    [II.A.1] (Reference #2, Ch.5—B.1)

4. a; In some companies the project selection group assigns some projects to Six Sigma teams and other projects to teams using other methodologies. For example, problems involving extensive data analysis and improvements using designed experiments would likely be assigned to a Six Sigma team, while a process improvement involving waste reduction might be assigned to a lean manufacturing team. [II.A.2] (Reference #2, Ch.1—A.2)

5. b; A more experienced cross-functional Six Sigma team will likely be assigned a project with a high level of technical complexity and stakeholder risk. [II.A.2] (Reference #3, II.A.2)

6. a; It is common to require that project proposals include precise statements of the problem definition and some preliminary measures of the seriousness of the problem, including its impact on the goals of the organization. For some managers,

these will be the criteria that define which projects to start first based on which ones save the most bottom-line dollars ($\overline{\overline{S}}$). [II.A.2] (Reference #2, Ch.4—A.1)

7. c; *Communities*—Stakeholders are those who have a vested interest in the process and/or its products and outputs. Generally, stakeholders of an organization include customers, suppliers, employees, investors, and communities. Stakeholder interest and involvement with the process may change over time depending on economic, contractual, and other influences. [II.A.3] (Reference #2, Ch.4—A.5)

8. b; *Employees*—Stakeholders are those who have a vested interest in the process and/or its products and outputs. Generally, stakeholders of an organization include customers, suppliers, employees, investors, and communities. Stakeholder interest and involvement with the process may change over time depending on economic, contractual, and other influences. [II.A.3] (Reference #2, Ch.4—A.5)

   Internal failure costs are costs incurred to remedy defects discovered before the product or service is delivered to the customer, and employees are the most likely to experience the direct effects of internal failure costs. [I.E.2] (Reference #1, Ch.8—Cost of Poor Quality)

9. a; *End users*—external customers who purchase products/services for their own use. [II.A.3] (Reference #1, Ch.15—Customer Identification)

10. d; *Suppliers*—It begins with defining the process of interest and listing on the right side the outputs that the process creates that go to customers, who are also listed. Suppliers and what they provide to enable the process (the inputs) are similarly shown on the left side. [II.A.4] (Reference #1, Ch.1—Terms, Concepts, and Principles)

11. c; *SIPOC*.

    When to use:

    - At the beginning of a project, to help define the important elements of the project
    - When it is not clear what the process inputs are, who supplies them, what the outputs are, or who the customers are
    - When there are many suppliers, inputs, outputs, and/or customers

    [II.A.4] (Reference #6, Ch.5—The Tools—SIPOC)

12. d; *SIPOC*—When identifying the process, it is important to recognize that processes usually affect multiple departments and organizations. Crossing functional areas (departments) and organizations (suppliers, intercompany)

can add challenges to an improvement project. The first step in recognizing the challenges is to understand the organizations and functional areas involved with the process. As noted, the SIPOC diagram can help in identifying these organizations and functional areas as process suppliers and customers. [II.A.4] (Reference #3, II.A.4)

13. b; The process of effectively integrating and managing components of the supply chain. [II.A.5] (Reference #1, Appendix C)

14. c; SCM is a set of approaches utilized to efficiently integrate suppliers, manufacturers, warehouses, and stores so that merchandise is produced and distributed at the right quantities, to the right locations, and at the right time in order to minimize system-wide costs while satisfying service level requirements. [II.A.5] (Reference #1, Ch.13—Supplier Relationships)

15. a; *Project scope* provides project limitations in terms of budget, time, and other resources. [II.B.1] (Reference #2, Ch.6—C.1; Reference #3, II.B.1)

16. c; A *problem statement* is a statement of fact as observed by the customer. It includes what, who, when, where, and how many. It also includes the goal of accomplishment or success factors. [II.B.1] (Reference #3, II.B.1)

17. a; A formal *communication plan* has several important components, one of which is establishing a *communication protocol* that consists of what, who, where, when, how, and how frequently communications occur. [II.B.2] (Reference #3, II.B.2)

18. b; Communication is key to the outcome of a project, be it a simple or complex project. A simple project may require daily or weekly meetings with project members meeting face-to-face or through virtual media to update the status of the project and discuss any new risks encountered.

    A complex project, however, may require a *formal communication plan*. This means setting up a communication protocol (what, who, where, when, how, and how frequently), escalation protocol, escalation threshold (when to escalate), communication effectiveness verification, and so on. Adding to the challenge is a project team that is spread across geographies (distance and time zones), has different cultures, and has a varying level of infrastructure. [II.B.2] (Reference #3, II.B.2)

19. b; A *work breakdown structure* (WBS) consists of activities identified from start to finish. The WBS is developed by the project team members by brainstorming. [II.B.3] (Reference #3, II.B.3)

20. c; A *Gantt chart* is used to graphically display a project's key activities and the duration associated with those activities. In addition to showing the key activities, the Gantt chart also shows the task milestones and the task

relationships between predecessor and successor tasks. [II.B.3] (Reference #2, Ch.7—D—Management and Planning Tools)

21. a; Too often, operators use the tools and processes listed in this book as independent events to satisfy some issue of the moment. Using project planning, you should look at the entire system and strive for synergy as you use communication and the tools and processes to contribute to continual improvement. Project planning becomes the tracking system to ensure that all the elements, tools, processes, communications, and so forth are brought together as a whole system for doing the work that you do in the shop.

    Effective project planning requires skills in the following areas:

    - Information processing
    - Communication
    - Resource negotiations
    - Securing commitments
    - Incremental and modular planning
    - Assuring measurable milestones
    - Facilitating top management involvement

    At times operators may become part of project plans that are developed in the shop. Working with such plans should help ensure the successful demonstration of continual improvement and the satisfaction of customers' wants and needs. [II.B.3] (Reference #4, Tools—Project Planning)

22. d; An *affinity diagram* organizes a large number of ideas into categories. This method taps the team's creativity and intuition. It was created in the 1960s by Japanese anthropologist Jiro Kawakita. [II.B.4] (Reference #6, Ch.5—The Tools—Affinity Diagramming)

23. d; The *activity network diagram* displays the sequential order in which tasks are carried out in a project. The diagram illustrates the efficiency of the schedule for the entire project. It can also reveal any potential scheduling and resource problems in the sequence planned. A main benefit of the arrow diagram is the calculation of the *"critical path"* of the project. This is the flow of critical steps in which a delay in any of the steps will affect the duration of the entire project. A project manager looks at the critical path closely to ensure that resources are adequately planned to help make up for any delays in the project. [II.B.4] (Reference #2, Ch.7—D; Reference #3, II.B.4)

24. a; A *matrix diagram* is typically used to discover and illustrate relationships between two groups of items. In the figure below, the two groups are the units of a training

course and the objectives of the course. The items in one group are listed across the top of the chart, and the items in the other group are listed down one side.

		Unit								
		1	2	3	4	5	6	7	8	9
Objectives	Review basics	⊙		⊙	○			Δ		
	Math skills	○	Δ	⊙			⊙	○		
	Communication skills						○		Δ	
	Attitude/motivation								Δ	
	Sketching			⊙			○			
	Ohm's law				⊙	○	Δ			
	Kirkoff's law					⊙	○			
	Thevinev's law						Δ		○	
	Heisenberg's uncertainty principle									

⊙ = Strong relationship ○ = Moderate relationship Δ = Weak relationship

Conclusions: Thevinev's law and communication skills covered only weakly
Attitude/motivation barely covered
Heisenberg's uncertainty principle not covered
Unit 2 contributes very little toward objectives
Unit 9 contributes nothing toward objectives

The team examines each square in the matrix and enters one of three symbols or leaves it blank depending on the relationship between the items in the row and column represented by the square. The most conventional symbols are shown in the example, although letters and numbers are sometimes used. The team then examines the completed matrix and discusses possible conclusions. [II.B.4] (Reference #2, Ch.7—D)

25. c; Phase reviews (tollgates) are completed to ensure that the project follows the necessary structure and deliverables are met. For this question, the correct answer "When the team has identified performance gaps and narrowed the focus of the project" indicates that the team is concluding the *measure* phase and is ready to present to the project sponsors the magnitude of the performance gaps as well as a revised problem statement(s). [II.B.5] (Reference #3, II.B.5)

26. c; As an example, recognizing the need for a Six Sigma project, building stakeholder consensus, and obtaining sponsor approval may likely be a milestone. Setting up a data collection system could be another milestone. The project manager and the team have to recognize the appropriate stages of the project to plan reviews. The purpose of this periodic review is to ensure that the project follows the necessary structure, deliverables are met, risks are anticipated and mitigated, project cost and time are closely monitored, and customers and other stakeholders are not negatively impacted. [II.B.5] (Reference #3, II.B.5)

Part III

Measure Phase

(30 questions)

QUESTIONS

1. Another term for the centerline of a control chart is the:

   a. median.

   b. mean.

   c. mode.

   d. margin.

Questions 2 and 3 are based on the following information.

A Six Sigma project team has been assigned a project focused on reducing cycle time. A sample of 10 data points has been collected:

10, 15, 12, 8, 20, 6, 9, 11, 21, 10

2. If the project team wanted to calculate the average of the set of values, which of the following is the correct value?

   a. 12.2

   b. 10

   c. 10.5

   d. 11.4

3. If the project team wanted to know the middle value, which of the following is the correct value?

   a. 12.2

   b. 10

   c. 10.5

   d. 11.4

4. Measures of dispersion are used to understand the spread or variation in a process. Which of the following uses the extreme values of minimum and maximum to describe variability?

   a. Interquartile range

   b. Standard deviation

   c. Range

   d. Coefficient of variance

5. A call center has recently experienced an increase in customer complaints regarding lengthy hold times. To understand the variability experienced by the customer, a sample of hold times in minutes was collected:

   5, 10, 15, 22, 8, 25, 22, 21, 18, 15

   The assigned project team decided to use the standard deviation to understand the dispersion of the data. Which of the following is the standard deviation for the hold times?

   a. 6.7

   b. 20

   c. 45

   d. 8.2

6. A hospital recently initiated a Six Sigma project to address an increase in patient falls. In the *measure* phase the team reviewed 10 weeks of historical data on the number of falls per week and calculated basic statistics to understand the centeredness and variation in the falls per week.

Wk 1	Wk 2	Wk 3	Wk 4	Wk 5	Wk 6	Wk 7	Wk 8	Wk 9	Wk 10
4.0	2.0	3.0	2.0	0.0	0.0	1.0	1.0	2.0	2.0

   The measures of central tendency are:

   a. mean 2.0, median 2.0, mode 4.0.

   b. mean 2.5, median 2.5, mode 1.0.

   c. mean 1.7, median 2.0, mode 2.0.

   d. mean 1.8, median 1.0, mode 2.0.

7. Which of the following is an example of attribute data?
   a. Length
   b. Tensile strength
   c. Weight
   d. Cleanliness
8. Which of the following is a data collection method considered to have a low response rate?
   a. Survey
   b. Focus group
   c. Mystery shopping
   d. Face-to-face interview
9. A team has been formed to determine the reason for the decrease in customer satisfaction in its gourmet cupcake store. The team would like to better understand the reason for the decline in customer satisfaction. Which of the following is the next step for the team?
   a. Perform a capability analysis
   b. Conduct a survey
   c. Develop a data collection plan
   d. Create a sampling plan
10. A data collection plan is used during the ____________ phase of a Six Sigma project.
    a. define
    b. measure
    c. analyze
    d. improve
11. A sampling plan contains ________, ________, ________, and __________.
    a. confidence level, power of the test, sample size, variance
    b. confidence level, power of the test, sample size, covariance
    c. what, where, who, how many
    d. what, where, when, how many

12. Quantitative measurements can be gained from which of the following information gathered by a company's supplier relations department?

    a. Personal interviews with supplier personnel

    b. Verbal feedback from the purchasing agent

    c. Results from formal, ongoing supplier surveys

    d. E-mail comments from local distributors

13. The quality department is offering a three-hour course on statistical process control with a focus on capability analysis. Which of the following types of data was used to conduct this analysis?

    a. Quantitative data

    b. Attribute data

    c. Ordinal data

    d. Qualitative data

14. Which of the following is a data collection tool used when collecting qualitative data?

    a. Check sheet

    b. Time tracker

    c. Random sampling

    d. Face-to-face interviews

15. The team assigned to the gourmet cupcake store discovered that customers experienced inconsistency in the ordering process. To better understand this variation, the team decided to construct an XmR chart. Which of the following types of data is required to create an XmR control chart?

    a. Ordinal data

    b. Nominal data

    c. Continuous data

    d. Discrete data

16. A Six Sigma team designed a quality survey containing five questions. One of the questions has two responses: "Yes" or "No." Which of the following represents this type of data?

    a. Categorical

    b. Continuous

    c. Variable

    d. Nominal

17. A check sheet was designed to track the frequency of occurrence of the use of three colors of paint in an automotive paint shop. Which of the following best represents this type of data and the measurement scale?

    a. Qualitative, ordinal

    b. Qualitative, nominal

    c. Quantitative, interval

    d. Quantitative, ratio

18. A Six Sigma team has formed to address increasing wait times in a hospital's emergency room services. One of the types of data the team has decided to collect is customer feedback. Which of the following techniques would be the best method for collecting this data?

    a. Check sheet

    b. Checklist

    c. Survey

    d. Automatic data capture

19. A large insurance company is designing a new online application process. The assigned project team would like to gather input from existing and future customers. Which of the following would be the best method for collecting this data?

    a. Face-to-face interviews

    b. Focus groups

    c. Online survey

    d. Observations

20. A Six Sigma project team is in the *analyze* phase and is ready to collect data to determine what types of errors are occurring during the admission process in a local emergency room. A simple random sample of applications has been collected for review of missing or incorrect fields. Which of the following is the best data collection method for this review?

    a. Survey

    b. Focus groups

    c. Checklist

    d. Observations

21. A banking company implemented a Six Sigma strategy with a goal of improving performance over the next 3–5 years. An initial step was to obtain cycle time data from multiple sites. The data existed in the data warehouse but needed to be extracted through an ad hoc report. Which type of data collection method is best for obtaining this data?

    a. Check sheet

    b. Checklist

    c. Survey

    d. Automatic data capture

22. A Six Sigma project team is planning to conduct a survey to collect customer satisfaction information for service received while dining at the company's seafood restaurant chain. The population is somewhat hard to reach, and it is known that people visiting this restaurant are internet savvy and prefer to receive information via e-mail. Which of the following data collection tools should the team use?

    a. E-survey

    b. Telephone survey

    c. In-person survey

    d. Pen and paper survey

23. If a project team wants to assess the amount of appraiser variation in a gage R&R study, which of the following elements would the team be measuring?

    a. Repeatability

    b. Precision

    c. Reproducibility

    d. Accuracy

24. In a gage R&R study, *bias* is similar to which of the following?
    a. Linearity
    b. Precision
    c. Calibration
    d. Accuracy
25. Which of the following is defined as "the drift in the average over time"?
    a. Linearity
    b. Precision
    c. Calibration
    d. Accuracy
26. *Repeatability* in a gage R&R study is defined as which of the following?
    a. Variation in measuring equipment when measured by one appraiser in the same setting at the same time
    b. Variation in measurement when measured by two or more appraisers multiple times
    c. Accuracy of measurement at various points in the measuring range of the equipment
    d. Closeness of agreement between the average of one or more measured results and a reference value
27. Which of the following is a standard process specification error percentage for MSA?
    a. 30% to 35%
    b. 35% to 40%
    c. 40% to 50%
    d. 50% to 60%
28. Gage R&R is used for which of the following reasons?
    a. To understand the uncertainty of the measurement system
    b. To understand the accuracy of the measurement system
    c. To understand the precision of the measurement system
    d. To understand the variation of the measurement system

29. Which of the following results indicates the measurement system is acceptable?

Gage R&R Study—X-Bar/R Method		
Source	**VarComp**	**% Contribution (of VarComp)**
Total gage R&R	0.09350	7.12
Repeatability	0.04075	3.10
Reproducibility	0.05275	4.02
Part-to-part	1.21982	92.88
Total variation	1.31332	100.00

a. 92.88

b. 4.02

c. 3.10

d. 7.12

30. In a gage R&R study the _____________ is used to assess the sources of variation that are statistically significant.

a. ANOVA table

b. % contribution of variation

c. % study variation

d. % tolerance

SOLUTIONS

1. b; The *mean* is a measure of central tendency and is calculated as follows: (Sum of values)/(Number of values) = $\sum x/n$. The centerline of a control chart is most often calculated as an average. [III.A] (Reference #2, Ch.13—D.3; Reference #1, Ch.9—Improvement Tools)

2. a; The average, or mean, is 12.2. The *average*, or *mean*, represents a set of observations or data that are added together and divided by the total number of data points. This provides the central value of the observations. [III.A] (Reference #3, II.A)

3. c; The median is 10.5. The *median* is the central value of a set of observations or data arranged in ascending or descending order. If the number of values in the data set is an odd number, then the median is simply the value in the middle of the ordered data set. [III.A] (Reference #3, III.A)

4. c; The *range* is a measure of dispersion and is calculated by subtracting the lowest value in the data set from the highest value. Variance and standard deviation are more robust measures of dispersion. Range can be easily influenced by extreme outliers in the data set. [III.A] (Reference #3, III.A)

5. a; The standard deviation is 6.7. *Standard deviation* is a measure of how the numbers are distributed in the data compared with the normal curve. The area under the curve between any two points, expressed in standard deviation units (Z-scores), can be determined from statistical tables. The Z-score is the number of standard deviations that the measurement is from the mean and is calculated by the formula $Z = (x—\mu)/\sigma$. The standard normal distribution has mean = 0 and standard deviation = 1. [III.A] (Reference #3, III.E.1; Reference #4, Tools—Histogram)

   To calculate the standard deviation, first find the variance. To find the variance, subtract the average from each data point and sum the differences. [III.A] (Reference #3, III.A)

$$s^2 = \frac{\sum(x_i - \bar{x})^2}{n-1}$$

   $s2$ = Variance

   xi = Term in data set

   $\bar{x}$ = Sample mean

   $\sum$ = Sum

   n = Sample

	x	$\bar{x}$	$(x - \bar{x})$	$(x - \bar{x})^2$
X^1	5		11.1	123.21
X^2	10		6.1	37.21
X^3	15		1.1	1.21
X^4	22		5.9	34.81
X^5	8		8.1	65.61
X^6	25		8.9	79.21
X^7	22		5.9	34.81
X^8	21		4.9	24.01
X^9	18		1.9	3.61
X^{10}	15		1.1	1.21
Σ	161	16.1		404.9

$$s^2 = \frac{\Sigma(x_i - \bar{x})^2}{n-1} = \frac{404.9}{10-1} = 44.98$$

To calculate the standard deviation, take the square root of the variance. [III.A] (Reference #3, III.A)

$$s = \sqrt{\frac{\Sigma(x_i - \bar{x})^2}{n-1}} = \sqrt{44.98} = 6.7$$

6. c; The measures of central tendency—mean, median, and mode—are calculated as follows.

Wk 1	Wk 2	Wk 3	Wk 4	Wk 5	Wk 6	Wk 7	Wk 8	Wk 9	Wk 10
4.0	2.0	3.0	2.0	0.0	0.0	1.0	1.0	2.0	2.0

To calculate the *mean,* sum the values and divide by the number of values.
Mean = 4.0 + 2.0 + 3.0 + 2.0 + 0.0 + 0.0 + 1.0 + 1.0 + 2.0 + 2.0 = 1.7

To calculate the *median,* order the numbers based on numerical value and find the middle value. If there are two middle values, then add the two middle values together and divide by 2.

$$\text{Mean} = 4.0 + 3.0 + 2.0 + 2.0 + 2.0 + 2.0 + 1.0 + 1.0 + 0.0 + 0.0 = 2.0 + 2.0 + \frac{4.0}{2.0} = 2.0$$

The *mode* is the most frequently occurring value in a data set. First, order the numbers based on numerical value, and then find the number that occurs most frequently.

Mode – 4.0 | 3.0 | 2.0 | 2.0 | 2.0 | 2.0 | 1.0 | 1.0 | 0.0 | 0.0 = 2.0

[III.A] (Reference #3, III.A)

7. d; *Attribute data* is also known as *discrete data*. Discrete data result from counting the occurrence of events; for example, the number of red and green balls in a bucket. The attribute is color, and the types of color are red and green. In this case the attribute is whether the item or area is clean, or it is not clean. [III.B.1] (Reference #2, Ch.13—D.1)

8. a; *Surveys* are used when a large sample of data needs to be collected. Inputs are quick and often the first instinct from the input provider. Response rates are usually low, but the data are inexpensive to collect. The quality of the survey feedback can vary significantly based on how the survey was created. [III.B.1] (Reference #3, III.B.3)

9. c; A *data collection plan* is used to organize and provide rigor in the data collection process. Most data collection plans contain the following information: (1) what to measure, (2) how to measure, (3) when to measure, (4) where to measure, (5) who will measure, and (6) a sampling plan. [III.B.1] (Reference #3, III.B.1)

10. b; A data collection plan is initiated in the *measure* phase of a Six Sigma project. [III.B.1] (Reference #3, I.A; III.B.1)

11. d; A *sampling plan* is a section of the data collection plan containing the elements of what, where, when, and how many for the data to be collected. [III.B.2] (Reference #3, III.B.1—Table 9.1)

12. c; *Quantitative* measures can be obtained from a number of sources, including product data, corrective actions (quality), cost, and service (delivery). Surveys are designed to translate customer opinion into a quantitative number that can be used as a measure of satisfaction. [III.B.2] (Reference #1, Ch.16—Voice of the Customer; Reference #2, Ch.13—D.1)

13. a; *Quantitative data* are grouped into two types: continuous (also called *variables*) and discrete (also called *attributes*). *Continuous data* result from measurement on some continuous scale such as length, weight, temperature, and so on. These scales are called *continuous* because between any two values there are an infinite number of other values. [III.B.2] (Reference #2, Ch.13—D.1; Reference #3, III.B.2)

14. d; *Face-to-face interviews*. This method provides highly reliable data, but the data are expensive to collect. The interviewee should be willing to commit time and provide honest feedback. The interviewee's body language and emotions provide additional perspective to data collection that is not available when soliciting inputs through surveys. [III.B.2] (Reference #3, III.B.3)

15. c; The *chart of individuals* (XmR) is a pair of control charts used to study variable data that are not generated frequently enough for an $\bar{X}$ and *R* chart and that form a normal distribution. Continuous data in time order are used for this chart. Data are not subgrouped. Individual data points are plotted on one chart, and the

differences between successive points—the moving ranges—are plotted on the second. [III.B.2] (Reference # 2, Ch.21—A.3)

16. a; *Categorical* variables represent types of data that may be divided into groups, such as "Yes" or "No." Categorical data are qualitative and can be expressed in words like "tall," "short," "hot," "cold," "long," "wide," or in larger groupings such as "high," "medium," "low." Certain process measures are expressed only in qualitative data, such as consumer experience and perceptions. [III.B.2] (Reference #3, III.B.2)

17. b; Collecting data on three types of paint color such as "red," "blue," "yellow," is an example of *qualitative data* and a *nominal measurement scale.* [III.B.2] (Reference #2, Ch.13—D.1)

18. c; *Surveys* are used when a large sample of data needs to be collected. Inputs are quick and often the first instinct from the input provider. Response rates are usually low, but the data are inexpensive to collect. The quality of the survey feedback can vary significantly based on how the survey was created. [III.B.3] (Reference #3, III.B.3)

19. b; *Focus groups.* In this method, a group of individuals is assembled to obtain perceptions, thoughts, beliefs, and opinions about a product or service. It is similar to the face-to-face interview in that the interviewer may be able to obtain direct feedback. However, the interactions between the focus group participants may be skewed by the dominant participants. Having an experienced moderator will help obtain unbiased feedback. [III.B.3] (Reference #3, III.B.3)

20. c; A *checklist* has a standard set of items/questions that are required to be verified/answered for a process or product. As an example, an equipment technician uses a maintenance checklist to go over a list of items and confirm each item's fitness for use. [III.B.3] (Reference #3, III.B.3)

21. d; *Automatic data capture* is real-time data capture. This may be expensive to set up due to interfacing the hardware and software with the source where the data are generated. This prevents errors in the data and the associated cost of correcting the errors and prevents wrong decisions made from erroneous data. One of the main advantages of automatic data capture is the availability of the data when taking action. There is no time lag between data collection and availability. This may be important where the risks are high, such as in finance, healthcare, and security management. A classic example is alerting a customer when fraud has been detected on the customer's account. [III.B.3] (Reference #3, III.B.3)

22. a; *E-surveys* are sophisticated internet-based surveys. They can be structured with interactive features such as graphics or programming that triggers alternate questions based on previous responses. E-surveys can be more convenient and

even enjoyable for respondents, leading to higher response rates. They can also be carried out faster, provide almost instant analysis, cost less than traditional mail or phone surveys, and sometimes allow better targeting of respondents. Respondents can be contacted via e-mail, with a hyperlink to the website hosting the survey, or the survey can be located at the organization's website, with an invitation to visitors to complete it. [III.B.3] (Reference #6, Ch.5—The Tools—E-Survey)

23. c; *Reproducibility,* the variation in a measurement when measured by two or more appraisers multiple times. Reviewing reproducibility during the *measure* phase helps process owners understand the variability due to human inability to reproduce the same measurement trial after trial. [III.C.1] (Reference #3, III.C.1)

24. d; *Accuracy,* the closeness of agreement between the average of one or more measured results and a reference value. [III.C.1] (Reference #3, III.C.1)

25. c; *Calibration,* the drift in average measurements of an absolute value. [III.C.1] (Reference #3, III.C.1)

26. a; *Repeatability,* the variation in measuring equipment when measured by one appraiser in the same setting at the same time. Reviewing repeatability during the *measure* phase helps process owners understand the suitability of the equipment. [III.C.1] (Reference #3, III.C.1)

27. c; When conducting a *measurement systems analysis,* it is not uncommon to find an error of 40% to 50% in the measurement system. [III.C.1] (Reference #3, III.C.1)

28. d; A *gage R&R study* analyzes variation of a measurement system that uses an instrument or gage. This variation is compared to the observed total variation in the process. The main purpose of a gage R&R study is to ensure that measurement variation is low enough that process measurements truly reflect the process. It is impossible to know whether a product is within specifications or to work on reducing process variation if measurement variation is so great that it masks process variation. [III.C.2] (Reference #6, Ch.5—Repeatability and Reproducibility Study)

29. d; The results for the *total gage R&R* are used to determine if the measurement system is acceptable. The total gage R&R measures the measurement system based on *repeatability* (how consistent the operator was in measuring the same part over and over, also referred to as *within operator* error) and *reproducibility* (the amount of difference in measurement when different operators measured the same item, also referred to as *between operator* error). A result $< 9\%$ for the total gage R&R means the measurement system is acceptable. If the total gage R&R is greater than 9%, then the measurement system needs to be improved. [III.C.2] (Reference #2, Ch.14—E; Reference #3, III.C.2)

30. a; The ANOVA, or analysis of variance, table for a gage R&R study is used to assess which sources of variation are statistically significant. It is used to analyze the study as a designed experiment. The factors are instruments, operators, and samples. This analysis is done with computer software. The results will include interactions between instruments and operators. [III.C.2] (Reference #2, Ch.14—E)

Part IV

Analyze Phase

(37 questions)

QUESTIONS

1. Which of the following tools is used to reduce problems arising from poor housekeeping or organization?

   a. 5 Whys

   b. 8D

   c. 5S

   d. A3

2. The ABC Company is studying problems due to delays across a number of different activities. Each activity has been tested and runs perfectly on its own. These delays can most effectively be identified through:

   a. value analysis.

   b. brainstorming.

   c. prioritization matrices.

   d. just-in-time.

3. ______________ is a system of sending delivery signals from downstream to upstream activities where the upstream supplier does not produce until the downstream customer signals the need.

   a. Kanban/pull

   b. Point-of-use storage

   c. Visual controls

   d. 5S

4. ____________ is a process analysis tool used to understand the rate of customer demand.

   a. Processing time

   b. Takt time

   c. Lead time

   d. Cycle time

5. Which of the following FMEA versions looks at how a part actually is produced on the shop floor and the various effects that can arise?

   a. System

   b. Process

   c. Design

   d. Product

6. In creating a risk priority number, what are the three scales that are calculated together?

   a. Cycle time, occurrence, priority

   b. Severity, range, priority

   c. Cycle time, range, detection

   d. Severity, occurrence, detection

7. In creating the risk priority number, the three scales are calculated using which mathematical function?

   a. Add

   b. Subtract

   c. Multiply

   d. Divide

8. In the first step of a process FMEA—review the design/process—the team does which of the following?

   a. Uses a flowchart to identify the scope and to ensure that every team member understands it in detail

   b. Lists possible causes for every failure mode

c. Assigns a detection rating based on the ability to detect the occurrence of the failure mode after it has happened but before the customer is affected

d. Develops an action plan to improve the current controls or reduce the frequency of occurrence of the cause

9. Which of the following tools is the best choice in collecting data to identify potential causes during a root cause analysis?

   a. Design of experiments

   b. Customer survey

   c. Force-field analysis

   d. Taguchi method

10. The ______________ is a good tool for prioritizing the potential root causes of a problem.

    a. histogram

    b. control chart

    c. radar chart

    d. matrix diagram

11. Root cause analysis depends on an accurate map of the process under study. Another term for process mapping is:

    a. flowcharting.

    b. critical path.

    c. is/is not analysis.

    d. brainstorming.

12. A diagram that graphically illustrates the relationship between a given outcome and all the factors that influence the outcome is referred as a:

    a. cause-and-effect matrix.

    b. Ishikawa diagram.

    c. histogram.

    d. control chart.

13. When starting on a new 8D (8 Disciplines) process, what is the first step?

    a. Map the current process

    b. Initiate a root cause analysis

    c. Organize a small group

    d. Stop future problems

14. When an improvement team needs a problem-solving approach to identify, correct, and eliminate recurring problems, which of the following might they correctly choose?

    a. Taguchi method

    b. 8 Disciplines

    c. Correlation analysis

    d. Quality function deployment

15. Complete the following sentence: The difference between corrective action and preventive action is:

    a. the level of management involved.

    b. who discovers the defect.

    c. the timing of the problem.

    d. how the resolution is documented.

16. Which of the following is a common approach to effective corrective action?

    a. 8 Disciplines

    b. Hoshin Kanri

    c. Correlation Analysis

    d. Taguchi method

17. When designing a manufacturing process for preventive action, which of the following are the two major poka-yoke system approaches that can be applied?

    a. Corrective and preventive

    b. Alpha and beta

    c. Major and minor

    d. Control and warning

18. During a process planning effort, which of the following tools identifies progressively finer levels of actions needed to design new products or services with no errors?

    a. Force-field analysis

    b. SIPOC

    c. Tree diagram

    d. Gantt chart

19. Which of the following is accurate for the normal distribution?

    a. The mean of the sampling distribution of means is equal to the mean of the population.

    b. The variance of the sampling distribution is not equal to the variance of the population.

    c. The normal probability plot graphically shows that the data deviate markedly from a straight line.

    d. The Anderson-Darling test statistic will have a corresponding p value $< .05$.

20. A Six Sigma project team is analyzing cycle time for an underwriting application process. The target for cycle time is < 30 minutes per application. The team performed descriptive statistics, and the output provides a graphical distribution of the data:

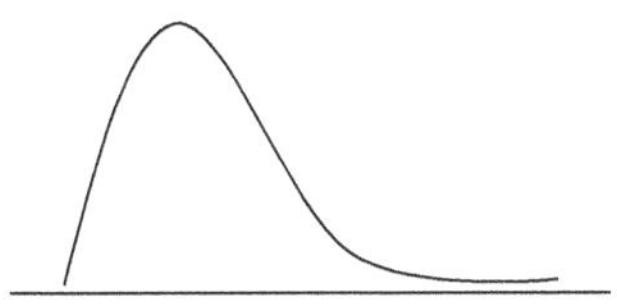

    Which of the following best describes the shape of this distribution?

    a. No skew

    b. Horizontal skew

    c. Positive skew

    d. Negative skew

21. Which of the following diagrams represents a bimodal distribution?

    a. 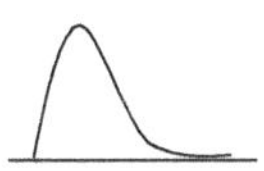

    b. 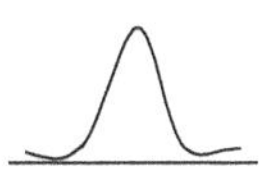

c. 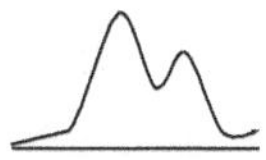

d. 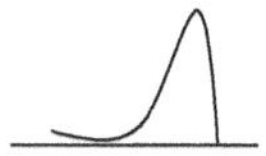

22. The binomial distribution is used to model which of the following types of data?

    a. Continuous data

    b. Discrete data

    c. Variable data

    d. Ordinal data

23. Which of the following formulas represents the normal distribution?

    a. $P(x) = le{-}1x$

    b. $P(x) = \frac{e^{-\frac{(x-\mu)^2}{2\sigma^2}}}{\sigma\sqrt{2\pi}}$

    c. $P(x) = \frac{n!}{x!(n-x)!}p^x(1-p)^{n-x}$

    d. $P(x) = \frac{e^{-1}l^x}{x!}$

24. When a system is affected only by common cause variation, that system:

    a. is optimized.

    b. is affected by external sources.

    c. meets a customer's quality specifications.

    d. is in control.

25. Which of the following describes when to use Hypothesis Testing:

    a. To understand the probability that results could have occured through random chance.

    b. To understand the likelihood that special cause variation is the source of the increase in cycle time.

c. To understand if there is a causal relationship between two variables.

d. To understand the relationship between multiple input variables and key output variables.

26. Which of the following tools is used to identify special cause variation?

 a. Run chart

 b. Pareto chart

 c. Control charts

 d. Histogram

27. Which of the following would indicate a strong positive linear relationship between two variables?

 a. .25

 b. .80

 c. .50

 d. .00

28. A Six Sigma project team wants to determine whether there is a linear relationship between the tenure of call center representatives and customer satisfaction. Which of the following would indicate there is a linear relationship between these two variables?

 a. Pearson correlation coefficient

 b. Coefficient of determination

 c. Coefficient of variance

 d. Kendal rank correlation

29. The correlation coefficient "*r*" is expressed as a number between ____________________ and ___________.

 a. 1.00, –1.00

 b. 0.50. –0.50

 c. 2.00, –2.00

 d. 1.50, –1.50

Questions 30–31 are based on the following information.

A large hospital identified a decline in safety scores across three locations. After engaging in root cause analysis, the project team discovered there was only one hospital with low scores, which were impacting the other two hospitals. The project team believed there was a strong positive relationship between safety scores and the length of time spent in safety training.

Model Summary

S	R-sq	R-sq(adj)	R-sq(pred)
6.47725	83.81%	83.10%	80.14%

Coefficients

Term	Coef	SE Coef	T-Value	P-Value
Constant	44.23	2.43	18.19	0.000
Time Training	0.6928	0.0635	10.91	0.000

30. Based on the output above, ________________ of the variation in safety scores is explained by time spent in training.

    a. 83.10%

    b. 80.81%

    c. 80.14%

    d. 69.28%

31. Which of the following statistics provides evidence that time spent in training is a significant driver of safety scores?

    a. $S = 6.44725$

    b. $R2$ (adj) = 80.14%

    c. $t = 18.19$

    d. p value < .05

32. Which of the following is a graphical component of regression analysis?

    a. Quality function deployment

    b. Scatter diagram

    c. Standard deviation

    d. Interrelationship digraph

33. Which of the following is a linear equation representing a straight line?

    a. $\hat{y} = a + bx$

    b. $\hat{y} = ax + b$

    c. $\hat{y} = a + bx^2$

    d. $\hat{y} = ax^2 + b$

34. A type I error occurs when the:

    a. null hypothesis is rejected when it is true.

    b. sample size is too small.

    c. null hypothesis is not rejected when it is false.

    d. power of the test is < .80.

35. Which of the following is representative of a two-tailed hypothesis test?

    a. H0: $\mu \leq 0$; H1: $\mu > 0$

    b. H0: $\mu \geq 0$; H1: $\mu < 0$

    c. H0: $\mu = 0$; H1: $\mu \neq 0$

    d. H0: $\mu > 0$; H1: $\mu \leq 0$

Use the following output from a one-sample *t*-test for questions 36–37.

```
Test of μ = 85 vs ≠ 85

Variable      N   Mean  StDev  SE Mean         95% CI      T      P
test scores  30  88.33   7.81     1.43  (85.42,91.25)  2.34  0.026
```

36. Which of the following is the null hypotheses?

    a. $\mu = 85$

    b. $\mu \neq 85$

    c. $\mu = 88.33$

    d. $\mu \neq 88.33$

37. Which of the following is the correct conclusion?

    a. Reject the null hypothesis

    b. Fail to reject the null hypothesis

    c. Accept the null hypothesis

    d. Fail to accept the null hypothesis

SOLUTIONS

1. c; 5S is a workplace organization method that can help improve the efficiency and management of operations. A process is impacted by its environment, as is the ability of personnel to respond to process change. Improvements in the general state of the work area, including access to hand tools, and so on, are an aid to process control. Especially critical here are the cleanliness, lighting, and general housekeeping status of any area where measurements are conducted since process control data are filtered through the measurement system. [IV.A.1] (Reference #2, Ch.2—B.1)

2. a; Waste can be identified by walking through the process from start to end and looking for any non-value-added activities. This is where value stream analysis is helpful. By closely designating every activity as either value-added or non-value-added, one should be able to identify wastes. Wastes like delays, travel distance, and excess inventory are addressed through tools such as value analysis, one-piece flow, Kanban, problem solving, SPC, and so on. [IV.A.1] (Reference #3, IV.A.1)

3. a; A *Kanban/pull* system sends delivery signals from downstream activities to upstream activities, where the upstream supplier does not produce until the downstream customer signals the need. [IV.A.1] (Reference #1, Ch.7—Process Improvement)

4. b; *Takt time* is the available production time divided by the rate of customer demand. Operating to takt time sets the production pace to customer demand. [IV.A.1] (Reference #1, Appendix C—Quality Glossary)

5. b; The process FMEA (the version operators should have available to them) looks at how the part actually is produced on the shop floor and the various effects that can arise. [IV.A.2] (Reference #4, Tools—Failure Mode and Effects Analysis [FMEA])

6. d; As the potential failure modes are identified by the cross-functional team, a determination of score needs to be developed for each of the three primary categories of severity, occurrence, and detection. [IV.A.2] (Reference #2, Ch.3—C.2)

7. c; Once the team has identified the three numbers for severity, occurrence, and detection (S-O-D) for a given failure mode, then the three numbers are multiplied together to create what is called the risk priority number (RPN). [IV.A.2] (Reference #2, Ch.3—C.2)

8. a; The first step in conducting a process FMEA is to use flowcharts to identify the scope and to make sure every team member understands it in detail. It is

also recommended that the team perform a walk-through of the process and understand the process steps firsthand. [IV.A.2] (Reference #3, IV.A.2t)

9. c; Collect data to help identify potential causes. Some of the techniques and tools that are useful include:

   - Brainstorming
   - Cause-and-effect diagram
   - Force-field analysis diagram
   - Flowchart

   [IV.B] (Reference #1, Ch.9—Improvement Tools)

   Force-field analysis is used to offer a counterpoint for every point that favors an issue or a decision. This tool is useful in understanding root causes by reviewing the forces "for" and "against." [IV.B] (Reference #3, IV.B)

10. d; Matrix diagrams and other prioritization tools can be used to prioritize root causes and corrective actions. The cause-and-effect diagram used to prioritize causes is a type of matrix diagram. During a root cause analysis, 30–50 causes may emerge from brainstorming. Teams often don't have the resources and time to investigate all of them. Prioritization tools can be used to focus on the top causes. [IV.B] (Reference #3, IV.B)

11. a; Some of the RCA methodologies in use include 8 Disciplines (8D) methodology, 5 Whys, Six Sigma–DMAIC (define, measure, analyze, improve, control), and drill deep and wide (DDW) (Ford Motor Company). Some techniques used to uncover the root cause include:

    - Value-added and non-value-added analysis
    - Comparison of the as-is processes and the should-be processes (flowcharts define processes)
    - Failure mode and effects analysis (FMEA)
    - Change analysis (evaluating changes made that could potentially affect the outcome)
    - Barrier analysis (errors in prevention—why did the inspection fail to capture the defect?)
    - Prediction analysis (why did the organization fail to predict the problem?)

    [IV.B] (Reference #2, Ch.19—B)

12. b; The cause-and-effect diagram graphically illustrates the relationship between a given outcome and all the factors that influence the outcome. It is sometimes called the *Ishikawa diagram* (after its creator, Kaoru Ishikawa) or the *fishbone diagram* (due to its shape). This type of diagram displays the factors that are thought to affect a particular output or outcome in a system. The factors are often shown as groupings of related sub-factors that act in concert to form the overall effect of the group. [IV.B] (Reference #1, Ch.10—Root Cause Analysis)

13. c; Use a team approach. Organize a small group (note we did not say that you have a team at this point—this is just a starting group) of people with knowledge of the process/product, authority to make changes on the shop floor, and skill in the required technical disciplines to solve the problem and implement corrective actions. [IV.B] (Reference #4, Tools—Problem Solving)

14. b; 8 Disciplines (8D) model—A problem-solving approach to identify, correct, and eliminate recurring problems. [IV.B] (Reference #1, Appendix C—Quality Glossary)

15. c; The difference between corrective action and preventive action is the timing of the problem. If the problem has already happened, the request is called a corrective action, and RCA focuses on the root cause that allowed the problem to happen in the first place. If the problem has not yet happened but is likely to happen, the request is called a preventive action, and RCA focuses on the root cause that could potentially allow the problem to happen. [IV.C] (Reference #2, Ch.19—B)

16. a; Some of the most successful attempts to solve problems have been accomplished through the use of a model or tools that outline the steps that should be followed in investigating and containing issues and fixing the problems so that they will not return. Unfortunately, many of us have seen situations where someone or even a group or team will fix a problem, only for the same issue to crop up again in a week, month, or year. The question is how do we permanently solve problems? One common approach to problem solving is called the 8 Discipline approach. [IV.C] (Reference #4, Tools—Problem Solving)

17. d; Two poka-yoke system approaches that lead to successful zero-defect systems are commonly utilized in manufacturing:

    1. Control approach: This takes the human element out of the equation and has a high capability of achieving zero defects:

        - It shuts down the process when an error occurs.
        - It keeps the "suspect" part in place when an operation is incomplete.

    2. Warning approach: The operator is signaled via some form of alarm (light/sound) to stop the process and correct the problem. This may be used when an automatic shutoff is not an option.

    [IV.D] (Reference #1, Ch.7—Process Improvement)

18. c; A tree diagram is a graphic representation of the separation of broad, general information into increasing levels of detail. The tool ensures that action plans remain visibly linked to overall goals, that actions flow logically from identified goals, and that the true level of a project's complexity will be fully understood. Tree diagrams are used in the quality planning process. The diagram begins with a generalized goal and then identifies progressively finer levels of actions needed to accomplish the goal. As part of process improvement, it can be used to help identify root causes of trouble. The tool is especially useful in designing new products or services and in creating an implementation plan to remedy identified process problems. [IV.D] (Reference #1, Ch.9—Improvement Tools)

19. a; The normal distribution is a continuous distribution used for variable data like measurements of length, mass, and time and is the one most frequently used by various professionals. The area under the curve between any two points, expressed in standard deviation units (Z-scores), can be determined from the statistical tables. The Z-score is the number of standard deviations that the measurement is from the mean and is calculated by the formula $Z = (x - \mu)/\sigma$. The standard normal distribution has mean = 0 and standard deviation = 1. [IV.E.1] (Reference #2, Ch.12—C)

20. c; The skewed distribution is unsymmetrical because a natural limit prevents outcomes on one side. The distribution's peak is off center toward the limit, and a tail stretches away from it. The positive, or right-skewed, distribution is one in which the tail is on the right side. Cycle time, where smaller is better, is often a right-skewed distribution. [IV.E.1] (Reference #6, Tools—Histogram and Other Frequency Distributions)

21. c; The bimodal distribution looks like the back of a two-humped camel. The outcomes of two processes with different distributions are combined in one set of data. For example, a distribution of production data from a two-shift operation might be bimodal if each shift produces a different distribution of results. Stratification often reveals this problem. [IV.E.1] (Reference #6, Tools–Histogram and Other Frequency Distributions)

22. b; The *binomial distribution* is used to model discrete data and should be applied in situations where each part has just two states, typically "good/bad" or "yes/no." [IV.E.1] (Reference #2, Ch.12—C)

23. b; The *normal distribution* is a continuous distribution used for variable data like measurements of length, mass, and time and is the one most frequently used by various professionals. [IV.E.1] (Reference #3, IV.E)

24. d; *Common cause variation* is variation inherent in the process. This type of variation is random and displayed between the upper and lower control limits. [IV.E.2] (Reference #1, Ch.9—Improvement Tools)

25. a; Hypothesis testing is an approach for analyzing data. It answers the question, "What is the probability that these results could have occurred through random chance?" The other side of that question is, "Is the effect we think we see in this data real?" Hypothesis testing is used when the question is about a large population but data are available only for a sample of the population. This approach is used to answer a wide variety of questions important in quality improvement, such as "Did the change we made to the process create a meaningful difference in the output?" or "Is customer satisfaction higher at location A than at our other locations?" [IV.E.2] (Reference #6, Ch.5—Hypothesis Testing)

26. c; The principal purpose of *control charts* is to recognize the presence of *special causes* so that appropriate action can be taken. While both special and common causes can be detected with statistical techniques, common causes are more difficult to isolate and remove. A process is considered to be in statistical control when *special causes* have been removed and only common causes remain. [IV.E.2] (Reference #3, IV.E.2)

27. b; Correlation analysis helps quantify a linear relationship between two variables. The analysis generates a correlation coefficient r that tells whether the relationship is linear, how strong it is, and whether the correlation is positive or negative. The correlation coefficient r lies between –1 and 1.

    - If r is close to zero, the two variables are not linearly related.
    - The closer r is to 1 or –1, the stronger the linear relationship of the two variables.
    - Positive r means that x is small when y is small and x is large when y is large.
    - Negative r means that x is small when y is large, and vice versa. [IV.F.1] (Reference #6, Ch.5—Correlation Analysis)

28. a; The Pearson correlation coeffcient tells whether the relationship is linear, how strong it is, and whether the correlation is positive or negative. [IV.F.1] (Reference #2, Ch.16—A.2)

29. a; The *correlation coefficient r* provides both the strength and the direction of the relationship between the independent and dependent variables. Values of r range between –1.0 and +1.0. When r is positive, the relationship between x and y is positive, and when r is negative, the relationship is negative. A correlation coefficient close to zero is evidence that there is no relationship between x and y. The strength of the relationship between x and y is measured by how close the correlation coefficient is to +1.0 or –1.0. [IV.F.1] (Reference #3, IV.F.1)

30. b; 83.81% is the R-sq value from the Model Summary. This is also called the Coefficient of determination, r^2. This tells us that 83.81% of the variaion in safety

scores is explained by time spent in training. This number, which is between 0 and 1, measures how well the data fit the line. If $r^2 = 1$, the line fits the data perfectly. As r^2 gets smaller, the line's fit becomes poorer, and predictions made from it will be less accurate. You can think of r^2 as the proportion of y's variation that is explained by the regression line. Because most data points don't fall exactly on the line, the rest of the proportion $(1 - r^2)$ is error. [IV.F.2] (Reference #6, Ch.5—The Tools—Regression Analysis)

31. d; Standard Statistical references define p-value as "the probability of obtaining a test statistic result at least as extreme or as close to the one that was actually observed, assuming that the null hypothesis is true." If the calculated p-value is greater than the significance level considered (often 1% or 5%), then the p-value is not significant enough to conclude that the null hypothesis is true. In this example, the time spent in training has a p value of 0.00 (<.05) indicating that time spent in training is a significant driver of safety scores. [IV.G] (Reference #2, Ch.15—G)

32. b; A *regression analysis* is a statistical tool used to find a model for the relationship between pairs (x, y) of numerical data. These pairs are plotted graphically in a scatter plot to provide a visual for determining if there is a linear relationship. The linear regression equation identifies the best straight line through a scatter diagram of the data. [IV.F.2] (Reference #6, Ch.5—The Tools)

33. a; A *simple regression* is used to describe a straight line that best fits a series of ordered pairs (x, y). An equation for a straight line, known as a *linear equation*, takes the form $\hat{y} = a + bx$ where

    - $\hat{y}$ = The predicted value of y, given a value of x
    - x = The independent variable
    - a = The y-intercept for the straight line
    - b = The slope of the straight line

    [IV.D.2] (Reference #3, IV.F.2)

34. a; There are two types of errors possible when formulating a conclusion regarding a population based on observations from a small sample. *Type I error* results when the null hypothesis is rejected when it is actually true. For example, incoming products are good, but labeled defective. This type of error is also called α *(alpha) error* and referred to as the *producer's risk* (for sampling). *Type II error* results when the null hypothesis is not rejected when it actually should have been rejected. For example, incoming products are defective, but labeled good. This type of error is also called β *(beta) error* and referred to as the *consumer's risk* (for sampling). [IV.E] (Reference #2, Ch.17—B.1)

35. c; If a null hypothesis is established to test whether a population shift has occurred in either direction, then a *two-tailed test* is required. A two-tailed hypothesis test is used whenever the alternative hypothesis is expressed as ≠. The allowable α error is generally divided into two equal parts. [IV.G] (Reference #3, IV.G)

36. a; A hypothesis is an assumption about a population parameter. All hypothesis tests have both a null hypothesis and an alternative hypothesis. A *null hypothesis,* denoted by H_0, represents the status quo and involves stating the belief that the mean of the population is ≥, =, or ≤ a specific value. The *null hypothesis* is believed to be true unless there is overwhelming evidence to the contrary. The alternative hypothesis, denoted by H_1, represents the opposite of the null hypothesis and holds true if the null hypothesis is found to be false. The alternative hypothesis always states that the mean of the population is <, ≠, or > a specific value. [IV.G] (Reference #3, IV.G)

37. a; A null hypothesis can only be rejected or fail to be rejected; it cannot be accepted because of a lack of evidence to reject it. [IV.G] (Reference #3, IV.G)

Part V

Improve and Control Phases

(26 questions)

QUESTIONS

1. A Japanese term that means gradual, unending improvement by doing little things better and setting and achieving increasingly higher standards is what?
   a. Poka-yoke
   b. Gemba
   c. Heijunka
   d. Kaizen

2. When a team must address a critical problem, they often schedule an intense, short time frame (typically 3–5 consecutive days) to apply the concepts and techniques of continual improvement (for example, to reduce cycle time, increase throughput, and reduce waste). What is this intense improvement activity called?
   a. Rapid cycle
   b. Kaizen blitz
   c. Kanban
   d. Six Sigma
3. Incremental improvement, or kaizen, is achieved by:
   a. frontline employees.
   b. top management.
   c. customers.
   d. middle management.
4. The individual attributed with making the practice of *kaizen* popular is:
   a. Kaoru Ishikawa.
   b. Masaaki Imai.

c. Eiji Toyoda.

d. Genichi Taguchi.

5. In which stage of the Deming cycle is a plan implemented on a trial basis?

   a. Study

   b. Test

   c. Do

   d. Act

6. In the ________ stage of the Deming cycle the team evaluates the effects of the implementation.

   a. check

   b. do

   c. plan

   d. act

7. The *plan* stage of the PDCA cycle includes which of the following activities?

   a. Maintain and standardize

   b. Set targets and schedules

   c. Plan implementation activities

   d. Select the next project

8. The PDSA improvement model was modified to PDCA by which of the following quality gurus?

   a. Dr. W. Edwards Deming

   b. Joseph M. Juran

   c. Philip B. Crosby

   d. Armand F. Feigenbaum

9. The time it takes to recoup the investment made in a project (the shorter the period, the better) is referred to as:

   a. payback period.

   b. recoup cycle.

   c. internal rate of return.

   d. return on investment.

10. A quality tool closely aligned with assessing the financial impact of poor quality is:

    a. statistical process control.

    b. the Shewhart cycle.

    c. cost/benefit analysis.

    d. quality function deployment.

11. Cost/benefit analysis is used in the _______________ phase of a Six Sigma project.

    a. define

    b. measure

    c. improve

    d. control

12. Control ________ provide a structured approach for the design, selection, and implementation of value-added control methods for the total system.

    a. panels

    b. charts

    c. indicators

    d. plans

13. The following steps would be used to create which document that ensures the correct monitoring and measurement of required actions and process outcomes?

    1. Specify those variables with direct or indirect impact to the remedy and the customer.

    2. Establish the control limits and standards for when to take action.

    3. Measure and set baselines for the different control variables.

    4. Specify the timing and location for measurements and determine the appropriate control chart.

    5. Assign and delegate those who will review and analyze monitoring and measurement results to identify when processes are out of control and diagnose the assignable causes.

    6. Put troubleshooting and corrective actions in place to restore the process quickly, incorporating adaptations to this document as part of ongoing improvement.

a. Flowchart

b. Control plan

c. Root cause analysis

d. Control chart

14. The ______________ combines necessary information into one document to help plan, monitor, control, study, and maintain your process.

a. dynamic control plan

b. measurement control plan

c. communication control plan

d. gage control plan

15. Which of the following is a control plan that provides for a written method to describe the system that controls the proper usage of the equipment to help ensure that measurement variation is as low as possible given the current set of conditions?

a. Gage

b. Dynamic

c. Frequency

d. Measure

16. The purpose of a control plan is:

a. to summarize the results of the capability analysis.

b. to document objectives of the project.

c. to explain how to control the workflow in a process.

d. to record information on sample size determination.

17. Which one of the following is an advantage of using a control chart?

a. Provide consistency between operators, shifts, and facilities

b. Provide a process for monitoring the voice of the customer

c. Provide performance results based on specification limits

d. Provide a method for analyzing qualitative data

18. Which of the following is true regarding control charts?

    a. Control charts can be used to monitor a process for the existence of common cause variation.

    b. The major components of a control chart are the lower specification limits, the centerline, and the upper specification limits.

    c. Control charts are essentially run charts to which two horizontal lines are added.

    d. Control charts alone can determine the source of problems.

19. Which of the following indicates on a control chart what a process should be capable of doing?

    a. Standard deviation

    b. Control limits

    c. Specification limits

    d. Variance

20. Which of the following is the formula is used to construct the upper control limits of the $\bar{X}$ and R chart?

    a. $\bar{\bar{X}} + A_2\bar{R}$

    b. $\bar{\bar{X}} - A_2\bar{R}$

    c. $\bar{X} + E_2\bar{R}$

    d. $\bar{X} - E_2\bar{R}$

21. What type of variation is displayed in the following individuals chart?

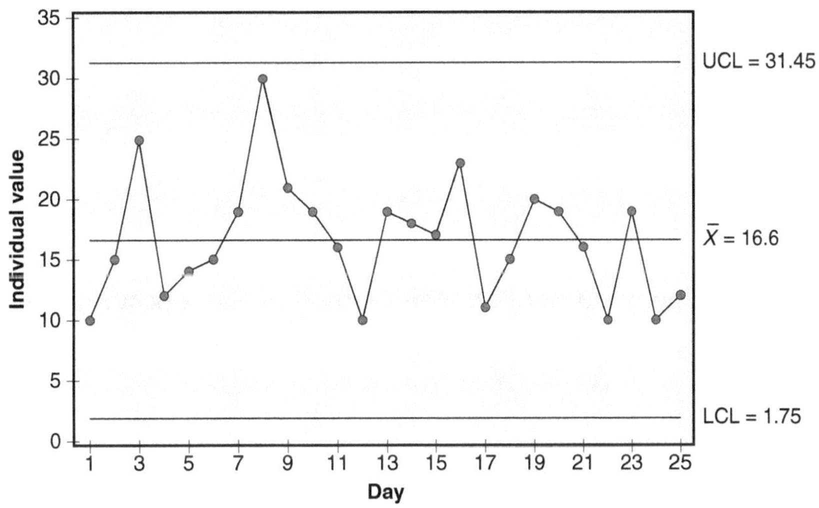

    a. Special cause

    b. Common cause

c. Simple cause

d. Relative cause

22. When intentional changes are made in a process, the operator should do which of the following?

    a. Conduct a root cause analysis

    b. Open an error record

    c. Gather voice of the customer data

    d. Update all related documents

23. The process during the *control* phase required to control document changes and communication to stakeholders is called:

    a. document control.

    b. stakeholder communication.

    c. process improvement.

    d. knowledge management.

24. A(n) ______________ is a step-by-step description of how to complete a task.

    a. quality management system

    b. deployment flowchart

    c. standard operating procedure

    d. operations checklist

25. Which one of the following would be used to help document work instructions?

    a. Matrix diagram

    b. Strategy map

    c. Flowchart

    d. Control plan

26. The work instruction document answers which of the following questions?

    a. Why is it done?

    b. How is it done?

    c. When is it done?

    d. Where is it done?

SOLUTIONS

1. d; *Kaizen*—incremental improvement; a Japanese term that means gradual, unending improvement by doing little things better and setting and achieving increasingly higher standards. Masaaki Imai made the term famous in his book *Kaizen: The Key to Japan's Competitive Success.*

2. b; *Kaizen blitz/event*—An intense, short-time-frame (typically 3–5 consecutive days) team approach to applying the concepts and techniques of continual improvement (for example, to reduce cycle time, increase throughput, and reduce waste). [V.A.1] (Reference #1, Ch.7—Process Improvement)

3. a; Incremental improvement, or kaizen, is achieved by the people who work on the process every day. By challenging themselves to improve quality and productivity without incurring significant capital investment, employees provide small, steady improvements in their work. Kaizen from day-to-day work is submitted by employees for review and approval. These small, gradual improvements are piloted and implemented across the organization. There needs to be management support and recognition of the efforts to institute kaizen improvements. Management needs to empower the employees to propose and make incremental changes to the products and processes. [V.A.1] (Reference #3, V.A.1)

4. b; *Masaaki Imai* made popular the practice of *kaizen*, a strategy for making improvements in quality in all business areas. Kaizen focuses on implementing small, gradual changes over a long time period. [V.A.1] (Reference #1, Ch.7—Process Improvement)

5. c; *Plan:*

   - Select project
   - Define problem and aim
   - Clarify / understand
   - Set targets / schedules
   - Inform and register the project
   - Solve / come up with most suitable recommendation

   Do:

   - Record / observe / collect data
   - Examine / prioritize / analyze
   - Justify / evaluate cost

- Investigate / determine most likely solutions
- Test and verify / determine cost and benefits
- Develop / test most likely causes

Check (study):

- Consolidate ideas
- Select next project
- Seek approval from management

Act:

- Plan installation / implementation plan
- Install / implement approved project / training
- Maintain / standardize

[V.A.2] (Reference #1, Ch.8—Improvement Techniques)

6. a; At the *do* stage, the team implements the solution. At the *check* stage, the team evaluates the effects of the implementation. At the *act* stage, the team takes appropriate actions as to whether to continue the cycle by going back to the *plan* stage or to standardize the learning to ensure sustainability of the actions. [V.A.2] (Reference #3, V.A.2)

7. b; *Plan:*

   - Select project
   - Define problem and aim
   - Clarify / understand
   - Set targets / schedules
   - Inform and register the project
   - Solve / come up with most suitable recommendation

   [V.A.2] (Reference #1, Ch.8—Improvement Techniques)

8. a; PDSA is a process intended to be a guiding light for the way we work and do our jobs. Building on the work of Walter Shewhart in the 1930s, *Dr. W. Edwards Deming* changed the improvement model PDCA (plan, do, check, act) to PDSA (plan, do, study, act). The latter model is the foundation for various processes such as advanced quality planning, problem solving, Six Sigma, and process improvement. [V.A.2] (Reference #3, V.A.2)

9. a; *Payback period.* The time it takes to recoup the investment made in a project: the shorter the period, the better. [V.A.3] (Reference #3, V.A.3)

10. c; Many things that are worked on throughout the shop can be classified into one of four categories: prevention costs, appraisal costs, internal failure costs, or external failure costs. On the other hand, organizations that have learned to use the cost/benefit analysis of quality cost, as called for in Six Sigma, are typically very surprised at the amount of waste that is being produced. By focusing on reducing prevention and appraisal costs, initial overall cost may rise; however, failure costs (internal and external) will slowly start to come down. [V.A.3] (Reference #2, Ch.1—A.1)

11. c; *Cost/benefit analysis* occurs in the *improve* phase once improvements have been identified. Cost of the improvement actions may involve design, development, deployment, sustaining people, equipment, material, environmental controls, software, communication, transport, security, and so on. These costs are compared with cost prevention or reduction of scrap, errors, defects, failures, accidents, violations (legal and regulatory), nonconformance, cost avoidance of repair, regrade, rework, recall, reinstall, and/or rehire. There are also costs that are difficult to measure, for example, loss of customer confidence and customer dissatisfaction. A typical cost/benefit analysis involves calculating both the applicable measurable costs that will be incurred and the benefits of the costs avoided. [V.A.3] (Reference #3, V.A.3)

12. d; Control plans provide a structured approach for the design, selection, and implementation of value-added control methods for the total system. [V.B.1] (Reference #4, Tools—Control Plan)

13. b; The following steps can be used to create a suitable control plan in order to ensure the correct monitoring and measurement of control subjects, and the appropriate actions to perform:

    1. Specify those variables with direct or indirect impact to the remedy and the customer.
    2. Establish the control limits and standards for when to take action.
    3. Measure and set baselines for the different control variables.
    4. Specify the timing and location for measurements and determine the appropriate control chart.
    5. Assign and delegate those who will review and analyze monitoring and measurement results to identify when processes are out of control and diagnose the assignable causes.

6. Put troubleshooting and corrective actions in place to restore the process quickly, incorporating adaptations to the control plan as part of ongoing improvement.

[V.B.1] (Reference #2, Ch.22—B)

14. a; A *dynamic control plan (DCP)* combines necessary information into one document to help plan, monitor, control, study, and maintain your process. Some of the documents include standard operating procedures (SOPs), control plans, FMEA, gage control plan, and quality planning sheets (QPSs). [V.B.1] (Reference #3, V.B.1)

15. a; A *gage control plan* is followed to look at the tools for monitoring and checking the process. Maintaining your tools is important to the safety and quality of your processes. The *gage control plan* can be a type of FMEA for the tools you use; it should look at maintenance, calibration, and proper handling of the instruments. The *gage control plan,* as the control plan, provides for a written method to describe the system that controls the proper usage of the equipment to help ensure that measurement variation is as low as possible given the current set of conditions. The *gage control plan* is not meant to replace the gage or test equipment instruction sheets, but to guide the operator in what to do if certain circumstances occur. [V.B.1] (Reference #3, V.B.1)

16. c; A *control plan* is a document summarizing all the key information that an operator needs to know to manage, control, and monitor the process. It includes the following elements: what to measure, when to measure, how to measure, how often to measure, and what to do and who to contact if something is not right. [V.B.1] (Reference #3, V.B.1)

17. a; Advantages of using *control charts* include (1) they allow operators to collect data at the process, (2) they increase yield by revealing and containing problems at the stage the problem is identified, (3) they help provide consistency between operators, and (4) they can be used to determine whether problems require local or management action. [V.B.2] (Reference #3, V.B.2)

18. c; A *control chart* and a *run chart* both have a central line for the average and are time ordered. Where the two charts differ is that the control chart has an upper line for the upper control limit, and a lower line for the lower control limit. [V.B.2] (Reference #6, Ch.5—The Tools—Control Chart)

19. b; The *control chart* is a graph used to study how a process changes over time. Control limits are determined from historical data using ±3 standard deviations from the mean and allow one to draw conclusions about whether the process variation is consistent (in control) or is unpredictable (out of control), affected by special causes of variation. [V.B.2] (Reference #6, Ch.5—The Tools—Control Chart)

20. a; $\bar{X}$ and R control charts show how the process average changes, along with corresponding changes in process variation. Control limits are calculated based on data from the process. The formulas for the $\bar{X}$ and R control limits are as follows:

    Upper control limit for the averages chart: $UCL_{\bar{X}} = \bar{\bar{X}} + A_2\bar{R}$

    A_2R Lower control limit for the averages chart: $LCL_{\bar{X}} = \bar{\bar{X}} - A_2\bar{R}$

    A_2R Upper control limit for the range chart: $UCL_R = D_4\bar{R}$

    Lower control limit for the range chart: $LCL_R = D_3\bar{R}$

    [V.B.2] (Reference #2, Ch.21—A.3 p. 390)

21. b; The *common causes of variation* are those that are inherent in the process and generally are not controllable by process operators. *Common cause variation* is also known as *natural variation* and refers to the many sources of variation within a process. *Common causes* reside in processes within statistical control and can be characterized by location (process average), spread (piece-to-piece variability), and shape (distribution) for predictability. *Common cause variation* is viewed as a stable process displaying random variation within the control limits. [V.B.2] (Reference #2, Ch.21—A.1)

22. d; When something changes in the process and a new desirable level is achieved, the operator should update all documents relating to that process. [V.B.3] (Reference #3, V.B.3)

23. a; Under the section for document control, organizations with ISO 9001 or equivalent management systems implemented have a process to control document changes and communication to stakeholders. *It is important to identify the appropriate stakeholders for change review, approval, and communication.* Sometimes the stakeholders may be outside the organization, for example, customers, suppliers, and the regulating body. Those with authority to make changes in the organization should also pay attention to retrieve obsolete communications and discard them. It is very easy in an organization to leave old information at various process locations if there is no system for controlled distribution and retrieval. This may cause major confusion and inconsistency in operations where one area is following the old process and another area is using the newly implemented changes. Change management also involves training personnel and mistake-proofing the system so that employees do not regress to the old ways. [V.B.3] (Reference #3, V.B.3)

24. c; When the process is developed, or updated, SOPs should be developed to ensure consistency in the process (whether manufacturing or service). This is one of the methodologies under the lean enterprise organization approach to standard work. An SOP is a step-by-step description of how to complete a task. [V.B.3] (Reference #3, V.B.3)

25. c; List of documents for Work Instructions. Work instructions can be documented as:

    - Written instructions
    - Checklists
    - Flowcharts
    - Photographs
    - Drawn pictures
    - Videos
    - Electronic screen shots
    - Electronic software-driven process steps

    [V.B.4] Reference #2, Ch.10—A)

26. b; Written Procedures and Work Instructions. Work instructions explain two other important aspects:

    - Who does what (personnel with specific skill set)
    - How it is done (step-by-step)

    [V.B.4] (Reference #2, Ch.10—A)

Section 2
Additional Practice Problems

(139 questions)

Section 2 contains 139 additional practice problems. These questions represent material from each of the five parts in the Six Sigma Yellow Belt Body of Knowledge. The questions are in a randomized order. Detailed solutions can be found in a section at the end of all the questions.

QUESTIONS

1. In a 300-person auto insurance organization forming a project team to determine ways to reduce the cycle time for processing policyholder claims, which of the following choices may represent the most effective size and composition of the project team?
   a. Four persons, all from the headquarters' claims processing function
   b. Three persons from the headquarters' claims processing (one from each of the three claims processing sub-functions) plus two persons representing field claims adjusters
   c. All 15 persons from the headquarters' claims processing function who have direct contact with policyholders
   d. One management representative from each of the 11 departments in the organization

2 Teams are generally formed in organizational settings by direction from a manager, or a(n):
   a. quality circle.
   b. sponsor.
   c. board of directors.
   d. audit committee.

3. Which of the following are responsibilities of a natural team?
   a. Set budgets, manage resources, and schedule work
   b. Assess other teams' work product and work closely with team members outside their department
   c. Make decisions affecting multiple departments and recommend new strategic initiatives
   d. Report to the same manager, share common goals, meet regularly
4. Which of the following is a characteristic of successful teams?
   a. Annual performance appraisals
   b. Total agreement on issues
   c. Established ground rules
   d. High member turnover
5. Which of the following is a characteristic of an effective project team member?
   a. Is independent of the department affected
   b. Prefers to work on tasks independently
   c. Creative and adept at problem solving
   d. Holds strongly to personal opinions
6. Which of the following is an initial question to be asked when establishing a team?
   a. What are the goals of this project?
   b. What is to be the final solution?
   c. When are the interim tasks to be completed?
   d. What type of control charts will be used to analyze the data?
7. Which of the following is a normal behavior during the *storming* stage of team growth?
   a. Arguing between members even when they agree on the real issue
   b. Ability to prevent or work through group problems
   c. Discussions of symptoms or problems not relevant to the task
   d. Attempts to determine acceptable group behavior and having to deal with group problems

8. During the first stage of a team's growth, the most appropriate leadership style for the team leader is to:

   a. resolve issues of power and authority.

   b. monitor progress and celebrate achievements.

   c. fully utilize team member's skills.

   d. provide clear direction and purpose.

9. Which of the following is the most usual progression through the four stages of team growth?

   a. The team generally spends longer in forming, then quickly progresses through storming, and alternates between norming and performing.

   b. The team moves quickly through forming to storming and norming, then achieves and remains in the performing stage.

   c. The pattern for each team is different through the four stages of team growth.

   d. The team moves quickly from forming and storming to norming, and only rarely achieves the full performing stage.

10. Which of the following flowchart formats is used to show major steps in a process and occasionally the next level of sub-steps?

    a. Deployment

    b. Top-down

    c. Opportunity

    d. Horizontal

11. Which of the following tools is most effective for determining how purchase orders are handled and who should handle them?

    a. Pareto chart

    b. Gantt chart

    c. Workflow diagram

    d. Deployment flowchart

12. The pattern reflected in a histogram when data values are subject to a natural limit is:

    a. skewed.

    b. bimodal.

c. bell-shaped.

d. symmetrical.

13. Which of the following tools is most effective in analyzing data collected on check sheets?

    a. Cause-and-effect diagram

    b. Workflow diagram

    c. Affinity diagram

    d. Pareto diagram

14. A work team has been assigned the responsibility of collecting data on the number of blemishes on cabinets in a laboratory scheduled for refurbishing. Which of the following tools would be most effective for this assignment?

    a. Check sheet

    b. *c*-chart

    c. *u*-chart

    d. Affinity diagram

15. Which of the following is the most effective way to choose project team members?

    a. Select all employees who can contribute something worthwhile to the improvement project

    b. Include only specialists who are experienced in details specific to the project being initiated

    c. Assign a combination of senior-level management and operations personnel to ensure top management commitment

    d. Include a small number of employees chosen from each area and worker level affected by the improvement

16. What is the ultimate goal of an effective supply chain?

    a. Customer loyalty

    b. Reduced costs

    c. Increased profits

    d. Speed to market

17. The purpose of a supply chain is to:

    a. Plan and manage a list of activities to successfully complete goals and deliverables.

    b. Develop collaborative approaches for delivering goods and services to customers.

    c. Build confidence that a supplier's product or service will fulfill its customer's needs.

    d. Place products and services timely and correctly in stores or at customer locations.

18. Which of the following tools would be used to categorize the outcome of a brainstorming session?

    a. Fishbone diagram

    b. Histogram

    c. Interrelationship digraph

    d. Affinity diagram

19. In reference to process management, removing the causes of an abnormal condition refers to_________, whereas _________ means changing the performance to a new level.

    a. performance, innovation

    b. control, improvement

    c. inspection, innovation

    d. variation, breakthrough

20. Which of the following control chart patterns is caused by special sources of variation that gradually affect the quality characteristics of a product or service?

    a. Cycle

    b. Shift

    c. Trend

    d. Mixture

Questions 21, 22, and 23 are based on the following information.

A hospital recently initiated a Six Sigma project to address an increase in patient falls. In the *measure* phase, the team reviewed 10 weeks of historical data on the number of falls per week and calculated basic statistics to understand the centeredness and dispersion of the data on falls per week.

Wk 1	Wk 2	Wk 3	Wk 4	Wk 5	Wk 6	Wk 7	Wk 8	Wk 9	Wk 10
4.0	2.0	3.0	2.0	0.0	0.0	1.0	1.0	2.0	2.0

21. The range for the 10-week period is:

    a. 4.0.

    b. 3.0.

    c. 3.5.

    d. 3.7.

22. The variance for the 10-week period is:

    a. 1.57.

    b. 1.25.

    c. 1.55.

    d. 1.28.

23. The standard deviation for the 10-week period is:

    a. 1.31.

    b. 1.28.

    c. 1.12.

    d. 1.25.

24. The operational definition in a data collection plan is defined as:

    a. measurement defined to describe the data to collect.

    b. measurement of the process improvement observed.

    c. measurement to achieve repeatable results from multiple observers.

    d. measurement description of the participants of the data collection effort.

25. The three elements of "what to measure" in the data collection plan are:

    a. measure, type of measure, sampling plan.

    b. measure, type of measure, operational definition.

    c. measurement or test method, stratification level, data collection method.

    d. measurement or test method, stratification level, sampling plan.

26. Which of the following is a type of quantitative data?

    a. Ordinal

    b. Nominal

    c. Continuous

    d. Categorical

27. A type of measurement scale that has a rational zero point for the scale is:

    a. ratio.

    b. ordinal.

    c. interval.

    d. nominal.

28. During the *define* phase of a Six Sigma project, a team gathered customer requirement information from customers through focus groups in three different locations in the country. The team concluded from their analysis that they needed further clarification on a couple of key factors. Given that the team has limited resources and a tight deadline, which of the following is the best method for obtaining specific information from the participants in the focus group?

    a. Face-to-face interview

    b. Telephone interview

    c. Internet survey

    d. E-survey

29. A local insurance company initiated a Six Sigma project where upper levels of leadership were directly involved. For this project the assigned team knew they would need to gather information at the executive level. Which of the following data collection techniques would be used to collect this type of information?

    a. Telephone interview

    b. Focus group

    c. Face-to-face interview

    d. Questionnaire

30. During a quarterly business meeting, an automotive company reviewed the focus group results presented by the Six Sigma team specific to the customer's experience with maintenance and servicing of its cars. The results provided significant insight into customer service requirements. In order to continue to

monitor and improve upon these results, leadership assigned the Six Sigma team the task of developing a data collection instrument that could be deployed to a large customer base on a quarterly basis. Which of the following data collection methods would a Six Sigma team recommend for collecting this information?

a. Telephone interview

b. Focus group

c. Face-to-face interview

d. Questionnaire

31. A process improvement team has conducted a measurement systems analysis (MSA). The team would like to know the difference between the absolute value and true value with respect to a standard master at various measurement points of the measuring range. Which of the following terms represents what the team is assessing?

    a. Linearity

    b. Bias

    c. Repeatability

    d. Stability

32. Which of the following best describes linearity in MSA?

    a. Closeness of repeated readings to each other

    b. Accuracy of measurement at various points of measuring range in the equipment

    c. Closeness of agreement between the average of one or more measured results and a reference value

    d. Drift in average measurements of an absolute value

33. The Six Sigma team has conducted a gage R&R study. The specific measure they are interested in is the variation in measurement between appraisers. Which of the following terms represents this variation?

    a. Repeatability

    b. Reproducibility

    c. Precision

    d. Accuracy

34. The purpose of a gage R&R study is to determine which of the following?

a. The part of variation in the data resulting from the variation in the measurement system

b. The part of variation in the data resulting from the shift in the mean of the measurement system

c. The part of variation in the process resulting from the data collected by the operator

d. The part of variation in the process resulting from the measures recorded by the operator

35. In a gage R&R study, the difference between using the ANOVA method and range method is that the ANOVA method provides which of the following?

a. Interactions between operators and samples

b. Interactions between each operator

c. Interactions between leadership and operators

d. Interactions between parts and tolerance

36. Consider the following two-way ANOVA table with interaction from a gage R&R study. What conclusions can be drawn from the part-to-operator interaction?

Gage R&R: Two-Way ANOVA Table with Interactions

Source	DF	SS	MS	F	P
Part	9	88.362	9.81799	492.290	0.000
Operator	2	3.1673	1.58363	79.406	0.000
Part * Operator	18	0.3590	0.01994	0.434	0.974
Repeatability	60	2.7589	0.04598		
Total	89	94.6471			

a. There is an interaction between part and operator.

b. There is not an interaction between part and operator.

c. There is a partial interaction between part and operator.

d. There is not a partial interaction between part and operator.

37. The eight types of waste are categorized as which of the following?

a. Rework

b. Value-added

c. Business value-added

d. Non-value-added

38. The purpose of FMEA is to:

    a. use process or design features to prevent the acceptance or further processing of nonconforming products.

    b. determine the set of components that could cause a failure in a process.

    c. understand the opportunities for failure and the impact of risks in a product or process design, and prioritize the risks.

    d. identify the ability of an item, product, or service to perform required functions on demand.

39. The ________ is a numeric assessment of risk assigned to a process, or steps in a process.

    a. occurrence number

    b. risk priority number

    c. severity number

    d. detection number

40. Which of the following is the likelihood or frequency at which the cause can happen?

    a. Severity

    b. Occurrence

    c. Detection

    d. Effectiveness

41. The ______________ is used to quantify and prioritize the impacts of causes (X) on effects (Y) through numerical ranking.

    a. Ishikawa diagram

    b. cause-and-effect matrix

    c. 5 Whys

    d. root cause tree

42. Which of the following is a root cause analysis tool that visually displays the process using shapes and symbols?

    a. Process flowchart

    b. Force-field analysis

    c. Ishikawa diagram

    d. Histogram

43. A binomial distribution should be applied when using which of the following types of data?

    a. Percent defective

    b. Red, yellow, orange, green

    c. Strongly disagree, disagree, neutral, agree, strongly agree

    d. Yes or no

44. Which of the following formulas represents the normal distribution?

    a. $P(x) = le^{-1x}$

    b. $P(x) = \frac{e^{-\frac{(x-\mu)^2}{2\sigma^2}}}{\sigma\sqrt{2\pi}}$

    c. $P(x) = \frac{n!}{x!(n-x)!}p^x(1-p)^{n-x}$

    d. $P(x) = \frac{e^{-1}l^x}{x!}$

45. Which of the following is a component of the central limit theorem?

    a. If the original population is distributed normally, the sampling distribution of means will also be normal.

    b. If the original population is distributed normally, the sampling distribution of means will not be normal.

    c. For large values of n, the distributions of the count X and the sample proportion p are approximately normal.

    d. The degrees of freedom when working with a single population variance is n–1.

46. A Six Sigma team has been formed to solve a problem in the discharge process at a local hospital. Recently, there has been an escalation in complaints from patients specific to the extremely long wait time from when the discharge orders have been submitted until the time they leave the hospital. The Six Sigma team analyzed the control charts and found three points outside the upper control limit. Which of the following is the type of variation in this process?

    a. Normal cause

    b. Common cause

    c. Natural cause

    d. Special cause

47. If you adjust a process in response to _________ variation, the result is usually more variation rather than less.

    a. unnatural cause

    b. assignable cause

    c. common cause

    d. special cause

48. Common causes reside in processes within statistical control and can be characterized by_________, _________, and _________ for predictability.

    a. location, variation, normal events

    b. location, spread, shape

    c. location, variation, unusual events

    d. location, spread, time

49. Correlation measures the _________ and _________ of the relationship between variables.

    a. strength, direction

    b. strength, predictability

    c. independence, dependence

    d. independence, interdependence

50. The _________ provides both the strength and the direction of the relationship between the independent and dependent variables.

    a. correlation of means

    b. correlation coefficient

    c. coefficient of determination

    d. coefficient of variance

51. Which of the following represents a strong positive correlation?

    a. .90

    b. .50

    c. .35

    d. .75

52. The ________ is a mathematical procedure for identifying the linear equation that best fits a set of ordered pairs by finding values for the y-intercept and the slope.

    a. least squares line

    b. weighted least squares

    c. least squares method

    d. estimated least squares

53. In the linear equation, b represents which of the following?

    a. y-intercept

    b. Slope of the line

    c. Independent variable

    d. Predicted value

54. Which of the following is used in regression to determine the percentage of the response variable variation that is explained by a linear model?

    a. r-multiple

    b. r

    c. r-squared adjusted

    d. r-squared

55. ________ is the probability of a type I error.

    a. Delta

    b. Alpha

    c. Beta

    d. Sigma

56. The ________ of a statistical test is the probability that it will correctly lead to the rejection of a false null hypothesis.

    a. power

    b. alpha

    c. sigma

    d. delta

57. *Kaizen* is defined as which of the following?

    a. Pursuit of excellence

    b. Continuous improvement

    c. Operational excellence

    d. Rapid improvement

58. Which of the following best describes the difference between *kaizen* and a *kaizen blitz/event*?

    a. Kaizen is defined as continuous improvement, and kaizen blitz is defined as a three-to-five-day event.

    b. Kaizen is defined as rapid improvement, and kaizen blitz is defined as a three-to-five-day event.

    c. Kaizen is defined as a process improvement team, and kaizen blitz is defined as continuous improvement.

    d. Kaizen is defined as a long-term project, and kaizen blitz is defined as a three-to-five-day event.

59. Which of the following is a concept described as a "basic belief that when quality becomes ingrained in the organization's culture and people, the quality of products and services will follow"?

    a. Kaizen

    b. Kaizen blitz

    c. Modular kaizen

    d. Kaizen ganza

60. Which of the following is the modified version of an improvement technique developed by Walter Shewhart and later modified by W. Edwards Deming?

    a. Plan-do-check-act

    b. Define-measure-analyze-improve-control

    c. Plan-do-study-act

    d. Define-measure-analyze-design-verify

61. The plan-do-check-act model is also referred to as which of the following?

    a. Juran cycle

    b. Deming cycle

c. Shewhart cycle

d. Crosby cycle

62. _________ involves calculating both the applicable measurable costs that will be incurred and the benefits of the costs avoided.

a. Return on investment analysis

b. Cost/benefit analysis

c. Payback period

d. Internal rate of return

63. The *payback period* is defined as which of the following?

a. A measure of the time it takes to recoup the investment made in a project; the shorter the period, the better

b. A measure to compare the profitability of investments; the higher, the better

c. A measure of the sum of the present values (PVs) of incoming and outgoing cash flows over a period of time; the higher, the better

d. A measure of the rate of return used in capital budgeting to measure and compare the profitability of investments; the higher, the better

64. A process improvement team is interested in measuring the length of time it would take to recoup the investment for the improvement recommendations in the *improve* phase of their project. Which of the following cost/benefit analysis methods would the team use?

a. Internal rate of return

b. Return on investment

c. Payback period

d. Net present value

65. A _________ is a written description of the systems for controlling part and process quality by addressing the key characteristics and engineering requirements.

a. quality plan

b. risk management plan

c. data collection plan

d. control plan

66. A rational subgroup is used in constructing which type of control chart?

    a. S chart

    b. $\bar{X}$ and R chart

    c. ImR chart

    d. X and S chart

67. A _________ is a sample set that is sufficient to determine common cause scenarios.

    a. rational group

    b. average subgroup

    c. median subgroup

    d. rational subgroup

68. A run chart is a line graph in which the vertical axis represents the _________ and the horizontal axis represents the _________.

    a. time, measurement

    b. cause, effect

    c. frequency, time

    d. measurement, time

69. A _________ is a step- by-step description of how to complete a task.

    a. reference document

    b. work instruction

    c. control plan

    d. checklist

70. A Standard Operating Procedure (SOP) is best described as:

    a. a high-level document that directly links to an organization's policies.

    b. a list of steps as a job aid for completing the task.

    c. legal document outlining the requirements for the task.

    d. an outline of the leadership responsibilities for monitoring task completion.

71. Six Sigma is a term used to describe the overall concept of:

    a. defect containment.

    b. process redesign.

    c. continual improvement.

    d. statistical process control.

72. Statistically speaking, _________ is a term indicating to what extent a process varies from perfection.

    a. alpha

    b. sigma

    c. omega

    d. beta

73. The quality concept of Six Sigma was initiated by which company in the 1980s?

    a. International Business Machines

    b. General Electric

    c. AT&T

    d. Motorola

74. Which of the following is a recognized value of using the Six Sigma approach for continuous improvement?

    a. Ensuring that improvement efforts are aligned with client and stakeholder needs

    b. The use of the PDCA cycle for process development

    c. Benchmarking to identify opportunities for improvement

    d. Containment of defective product for analysis and rework

75. A _________ is the series of activities that an organization performs.

    a. kaizen blitz

    b. value stream

    c. poka-yoke

    d. control plan

76. Which of the following uses simple graphics and icons to illustrate the movement of material, information, inventory, work-in-progress, operators, and so on through a process?

    a. Value stream map

    b. Deployment flowchart

    c. Cause-and-effect diagram

    d. Rapid cycle PDCA

77. Which of the following is "a systematic approach in identifying and eliminating waste (non-value-added activities) through continuous improvement by flowing the product at the pull of the customer in pursuit of perfection"?

    a. PDSA

    b. Gemba

    c. Lean

    d. Six Sigma

78. Which of the following describes "a production strategy promoted by Toyota, and now applied to many organizations, that strives to improve business return on investment by reducing in-process inventory and associated carrying costs"?

    a. Just-in-time

    b. Heijunka

    c. Quality circles

    d. Return on investment

79. Which level of Six Sigma Belts is responsible for facilitating Six Sigma methodologies, statistical tools, basic financial tools, change management, risk assessment, and project management activities?

    a. White

    b. Yellow

    c. Green

    d. Black

80. Which of the following Six Sigma roles is responsible for coaching and mentoring Black and Green Belts?

    a. Process owner

    b. Master Black Belt

c. Yellow Belt

d. Project sponsor

81. Which of the following Six Sigma stakeholders is responsible for interfacing with senior management concerning a Six Sigma project?

a. Yellow Belt

b. Team facilitator

c. Black Belt

d. Project champion

82. The role of the Yellow Belt in a Six Sigma project is usually as a:

a. team member.

b. subject matter expert.

c. process owner.

d. statistical specialist.

83. Which of the following is the recommended or usual number of core team members for a Six Sigma project team?

a. 13–15

b. 5–9

c. 8–12

d. 1–3

84. A _________ team enables people from all over the globe to meet via teleconference and internet tools.

a. natural

b. corrective action

c. virtual

d. work group

85. Which of the following types of teams tends to share leadership equally among group members, holding all members responsible for the output of the team?

a. Process improvement

b. Kaizen blitz

c. Corrective action

d. Self-directed

86. A senior manager has requested that a Six Sigma team be chartered to study and improve a complex process that spans several departments within the local facility. Which of the following types of teams should be commissioned?

    a. Virtual

    b. Cross-functional

    c. Natural

    d. Corrective action

87. During which stage of team development do members reconcile competing loyalties and responsibilities?

    a. Adjourning

    b. Storming

    c. Norming

    d. Forming

88. As team members become more comfortable with each other, and as they better understand the work and what is expected of them, the team enters which stage of team development?

    a. Norming

    b. Storming

    c. Adjourning

    d. Performing

89. Reviewing lessons learned, assessing the achievement of the outputs and outcomes intended, completing documentation, recognizing the team's efforts, and celebrating are activities during which stage of team development?

    a. Performing

    b. Adjourning

    c. Storming

    d. Norming

90. A significant activity to be performed early in the *forming* stage of team development is:

    a. determining acceptable team behavior and how to deal with team problems: establishing team ground rules.

b. questioning the wisdom of those who selected the project and appointed the other members of the team.

c. experimenting with ways to raise and discuss differences of opinion effectively.

d. preventing or working through group problems.

91. Which of the following is a reason why multivoting is preferable to straight voting when making a decision?

    a. It can be performed by an individual using a decision matrix rather than involving a whole team.

    b. It allows an item that is favored by all, but not the top choice of any, to rise to the top.

    c. Multivoting must be performed over a sequence of votes, where straight voting is performed only once.

    d. Hidden agendas among team members are less likely to influence the results using multivoting.

92. __________ is a structured process that identifies and ranks major problems or issues that need addressing.

    a. Quality function deployment

    b. Cycle time charting

    c. Nominal group technique

    d. Top-down flowcharting

93. Generating a large number of ideas surrounding an issue or a problem is performed during which stage of brainstorming?

    a. Evaluation

    b. Forming

    c. Opening

    d. Creative

94. For which of the following reasons might a Six Sigma team wish to use the brainwriting tool rather than the more direct brainstorming approach?

    a. If there is a limited time for the team to generate ideas

    b. When participants might feel safer contributing ideas anonymously

    c. If the sponsor has asked the facilitator to identify which participants contribute which ideas

    d. When the list of ideas generated needs to be included in the meeting minutes

95. Which of the following is an effective way to ensure that the key decisions made and actions agreed to in team meetings are formally recorded?

    a. Policies

    b. Standards

    c. Minutes

    d. Ground rules

96. When calling a team meeting, it is important to set an agenda for which of the following reasons?

    a. To identify the members of the Six Sigma team

    b. To set measurements for completing the project

    c. To reserve the meeting space and presentation materials

    d. To describe how the outcome can be accomplished

97. When should the team review their project status and communicate with major stakeholders?

    a. At the end of each major step in their project or work plans

    b. Before each team meeting, to ensure they are on track

    c. After each team meeting, to verify accuracy of any decisions made

    d. Only at the end of the project, to share results and lessons learned

98. An important function to be completed before starting a meeting is selecting the conference room, technology, room setup, and so on. Which of the following is the term used for this process?

    a. Kaizen

    b. Logistics

    c. Poka-yoke

    d. Dynamics

99. Which of the following basic quality tools is used to display what happens to one variable when another variable is changed?

    a. Cause-and-effect diagram

    b. Histogram

    c. Scatter diagram

    d. Flowchart

100. When collecting data on the frequency or patterns of events, problems, defects, defect location, or defect causes, which of the following tools should be employed?

    a. Histogram

    b. Run chart

    c. L-shaped matrix

    d. Check sheet

101. When first learning to use the cause-and-effect diagram, many teams will use which of the following approaches to label the major "fishbones," or categories, on the diagram?

    a. 6 M's

    b. 5 Whys

    c. 8D

    d. 7 Tools

102. Run charts differ from control charts because they:

    a. use the Y-axis to measure time sequence.

    b. do not have statistical control limits to monitor variation.

    c. use specification limits rather than control limits.

    d. are used for larger populations of data.

103. Which of the following concepts describes costs incurred because things are not done right the first time, every time?

    a. Project outcome cost analysis

    b. Taguchi method

    c. Cost of poor quality

    d. Critical-to-quality analysis

104. Quality planning—creation of plans for quality, reliability, operations, production, and inspection—is an example of which *cost of quality* category?

    a. Appraisal

    b. Planning

    c. Control

    d. Prevention

105. The total number of defects divided by the total number of products produced in some time period (for example, per day) is calculated as:

    a. DPU.

    b. DPMO.

    c. sigma.

    d. cycle time.

106. When a task is taking longer than the customer or the downstream process expects, which of the following activities should the team employ to resolve the issue?

    a. Quality function deployment

    b. Cycle time reduction

    c. Takt time analysis

    d. Is/is not matrix

107. Methods used to identify customers include which of the following?

    a. Requirements and measures tree

    b. Value stream mapping

    c. Tracking a product to delivery

    d. Nominal group technique

Questions 108, 109, and 110 are based on the following information.

> Department leadership at a consolidated call center supporting a large county health center met with a broad base of community health providers, clients, and service partners to learn of their priority needs for using the call center. Prompt response time was a major external customer requirement, identified in the statement "I consistently wait too long to speak with a call center representative."

108. In what type of activity is the department leadership engaging during this meeting?

    a. Gap analysis

    b. Failure mode and effects analysis

    c. Cost of quality

    d. Voice of the customer

109. Which of the following identifies an individual or groups with an interest in an issue?

    a. Pareto diagramming

    b. Stakeholder analysis

    c. Process drill-down

    d. Measurement stratification

110. After gathering data from the community representatives, the department leaders realized that responding to customer calls within a specific amount of time was a high priority. What type of requirement is the call center putting in place to track and control this response time?

    a. Cost containment

    b. Stakeholder stage gate

    c. Critical-to-quality

    d. Senior management review

111. When selecting a Six Sigma project, which of the following is the most critical characteristic?

    a. Saving the most bottom-line dollars

    b. Reducing waste and excess inventory

    c. Receiving the highest value in multivoting

    d. Getting the highest rating on an employee survey

112. In general, problems involving extensive data analysis and improvements using designed experiments would likely be assigned to which type of improvement team?

    a. Lean

    b. Natural

    c. Kaizen

    d. Six Sigma

113. Which of the following is a likely candidate to be on a Six Sigma project selection team?

    a. Human resources manager

    b. Master Black Belt

c. Yellow Belt

d. Purchasing agent

114. During which phase of the Six Sigma improvement cycle is a project scoped?

a. Measure

b. Control

c. Plan

d. Define

115. The larger group of individuals who have a vested interest in the process or its products and services are referred to as which of the following?

a. Creditors

b. Managers

c. Stakeholders

d. Candidates

116. A Six Sigma team needs help on a specialized task within the process they are studying. What is the term used for the individual who can meet this team requirement?

a. Master Black Belt

b. Subject matter expert

c. Senior engineer

d. External consultant

117. If the Six Sigma improvement team wishes to gather direct feedback from someone who uses their product in their own home, which of the following would they interview?

a. End user

b. Distributor

c. Wholesaler

d. Process owner

118. There is usually one person who is responsible for the overall design and maintenance of a set of tasks that accomplish a particular outcome. This person

is often invited to be on the Six Sigma team or at least serve as a subject matter expert. This person is called the _________ _________.

a. Black Belt

b. end user

c. process owner

d. team facilitator

119. A Six Sigma Yellow Belt team member has been asked to develop a high-level description of a process, including the inputs and outputs. Which of the following tools should she employ?

a. Affinity diagram

b. PERT

c. Run chart

d. SIPOC diagram

120. Which of the following contains the correct listing of the components of a SIPOC diagram?

a. Suppliers, inputs, process, outputs, customers

b. Suppliers, investors, partners, owners, customers

c. Services, income, profits, outlays, costs

d. Services, investments, profits, opportunities, constraints

121. When defining a process, it is important to identify which of the following goes to the customers?

a. Deliverables

b. Supplies

c. Outputs

d. Inputs

122. When planning to study a process or system, it is very important to first identify the start and end points of the process. These points are referred to as the _________ of the process.

a. control limits

b. capability

c. specification limits

d. boundaries

123. Which of the following is used during a team project as an informal contract that helps the team stay on track with the goals of the organization?
    a. Charter
    b. Memorandum of understanding
    c. SOP
    d. Service level agreement

124. A project goal can be defined as which of the following?
    a. The quantitative data gathered to measure the current state of a process
    b. The point toward which management directs its efforts and resources
    c. The measurement used to gauge progress toward the solution
    d. The final score from the end-of-project evaluation

125. One of the most important activities before setting project goals is to:
    a. establish the communication plan.
    b. brainstorm solutions.
    c. gather baseline data.
    d. conduct a root cause analysis.

126. Which of the following activities should be performed during the first team meetings of the *define* phase of DMAIC?
    a. Brainstorm problem solutions
    b. Conduct root cause analysis
    c. Develop a voice of the customer survey
    d. Review the project charter

127. The purpose of the communication planning worksheet is to:
    a. identify people who will be involved in or affected by your project.
    b. establish the Gantt chart to schedule the critical path for project tasks.
    c. design the data gathering and reporting procedures for the project.
    d. plan the voice of the customer surveys to gather baseline data.

128. Which of the following describes an update protocol (what, who, where, when, how, and how frequently), escalation procedure, escalation threshold (when to escalate), and feedback effectiveness verification for a Six Sigma process team?

    a. DMAIC process

    b. Communication plan

    c. Interrelationship digraph

    d. Critical-to-quality exercise

129. Effective project planning requires several qualitative or behavioral skills. Which of the following is one of those skills?

    a. Communication

    b. Statistical process control

    c. Confrontation

    d. Siloed thinking

130. A project management planning tool by which a project is decomposed into tasks, subtasks, and units of work is called a:

    a. standard operating procedure.

    b. work breakdown structure.

    c. design of experiments.

    d. decision tree analysis.

131. A Six Sigma project team has been chartered to resolve a disagreement between technicians on the exact procedure for performing a task. Which of the following tools should the team employ to identify the correct tasks and sequence of activities?

    a. Customer focus group

    b. Lessons learned analysis

    c. Work breakdown structure

    d. Nominal group technique

132. The Six Sigma team is developing a chart to schedule and monitor their project tasks. Which of the following tools should they employ?

    a. PERT

    b. Radar chart

c. SIPOC

d. Gantt chart

133. Which of the following tools is used to discover and illustrate relationships between two groups of items?

a. Matrix diagram

b. Force-field analysis

c. Scatter diagram

d. Is/is not matrix

134. When the project team must identify causal influences among different stakeholders of a process under study, which of the following tools should they use?

a. Regression analysis

b. Interrelationship digraph

c. Cause-and-effect analysis

d. Process decision program chart

135. Which of the following is used to analyze and display subtopics and topics branching out from the main topic?

a. Arrow diagram

b. Top-down flowchart

c. Tree diagram

d. Box plot

136. Which of the following tools highlights key tasks, the time to accomplish the tasks, and flow paths (serial or parallel)?

a. Quality function deployment

b. Failure mode and effects analysis

c. Critical-to-quality tree diagram

d. Activity network diagram

137. Some organizations set up entry and exit criteria for every Six Sigma project stage to ensure that all the necessary due diligence is completed before moving to the next stage. There may also be significant milestones between stages. The

status of these criteria is usually reported to the project sponsor or leadership during which of the following?

a. Phase review

b. After action report

c. Project closeout review

d. Team charter evaluation

138. Which of the following is a term used to describe the periodic progress reviews conducted for Six Sigma projects?

a. Evaluation session

b. Tollgate

c. Periodic report

d. Audit

139. A Six Sigma project team has reached a point during the *define* phase where they must seek sponsor approval for additional resources. Which of the following should they request from the sponsor?

a. Root cause analysis

b. Activity network diagram

c. Milestone review

d. Project audit

SOLUTIONS

1. b; Select the team members. Keep the team tight. Typically, teams should have no more than five to seven members in addition to the team leader and coach. Ideally, team members should represent each area affected by the project and each level of employees affected. Sometimes, they represent different stages of the process under study. They can be of various ranks, professions, trades, classifications, shifts, or work areas. [II.B.1] (Reference #5, p. 2–15)

2. b; Sponsor. The Sponsor is an individual manager who reviews and supports team efforts. An effective sponsor has a number of responsibilities one of which is to determine resources for the team. [I.D.1] (Reference #5, p. 2–6)

3. d; A *natural team* (such as a work group, department, or function) is made up of persons who have responsibility for a specific process or function and who work together in a participative environment. Unlike the process improvement team, the natural team is neither cross-functional nor temporary. The team leader is generally the person responsible for the function or process performed within the work area. The natural team is useful in involving all employees in a work group in striving for continual improvement. Starting with one or two functions, successful natural teams can become role models for expansion of natural teams throughout an organization. A natural team example follows.

   The information technology (IT) department serves all the functions within Mars Package Delivery's 4000-person countrywide operations. The IT department's work units (technical system maintenance, application systems design and programming, data entry, computer operations, data output, and customer service—internal, technology help desk, and administration) function as an internal team. Selected representatives from each work unit meet weekly to review the IT department's performance and to initiate corrective and preventive actions. [I.D.1] (Reference #1, Ch.4—Types of Teams)

4. c; One of the underlying principles pertaining to launching most any team is:

   - The team must be given or define for itself the ground rules and schedules under which it will operate.

   [I.D.1] (Reference #1, Ch.6—Initiating Teams)

5. c; When selecting team members, consider each individual's personal experience with the process; interest in improvement methods; creativity and problem-solving abilities; knowledge of improvement tools; ability to think analytically; skills in meeting facilitation, conflict management, communication, documentation (writing procedures), data collection, and analysis; and commitment to serving customers. Consider the individual's ability to learn

and to work collaboratively, level of initiative, and openness to new ideas. [I.D.1] (Reference #5, p. 2–15)

6. a; According to B. W. Tuckman, teams typically go through the stages of *forming, storming, norming, performing,* and *transitioning* (or adjourning).

   Stage 1: Forming:

   1. Team members getting to know each other
   2. Group is immature
   3. Sense of belonging to the group
   4. Taking pride in membership with the group
   5. Trying to please each other
   6. May tend to agree too much on initial discussion topics
   7. Not much work is accomplished
   8. Members' orientation on the team goals
   9. Members understand the roles and responsibilities

   [I.D.2] (Reference #3, I.D.2)

7. a; *Storming* includes these behaviors:

   - Arguing among members even when they agree on the real issue
   - Being defensive and competitive; splitting into factions and "choosing sides"
   - Questioning the wisdom of those who selected the project and appointed the other members of the team
   - Establishing unrealistic goals; expressing concern about excessive work
   - Perceiving a "pecking order"; creating disunity, increased tension, and jealousy
   - Withdrawing—whether literally dropping out of the team or simply withdrawing psychologically

   [I.D.2] (Reference #5, p. 6–6)

8. d; Leading a team through *forming.* To build trust and confidence during the forming stage, the team leader should:

   - Help members get to know each other
   - Provide clear direction and purpose

- Involve members in developing plans, clarifying roles, and establishing ways of working together
- Provide the information and structure the team needs to get started
- Help members answer the following questions:
  - "What will be expected of me?"
  - "Do I belong here?" "Do I *want* to belong here?"
  - "What's in it for me?"
  - "What's the purpose?"

[I.D.2] (Reference #5, p. 6–4)

9. c; The pattern is different for each team. Team members' attitudes depend on both the speed of progress and the resistance or encouragement they receive from their sponsor and their departments. [I.D.2] (Reference #5, p. 6–9)

10. b; A *top-down flowchart* shows the major steps and the first layer of sub-steps. It is used to understand or communicate the major steps of a process plus the key activities that constitute each step, when you want to focus on the ideal process, or when you can't see the forest for the trees. [I.E.1] (Reference #6, Ch. 5—Top-down Flowchart)

11. d; A *deployment flowchart* is a detailed flowchart that also shows who (which person or group) performs each step. This type of map is also referred to as a *cross-functional* or *swim lane* flowchart. [I.E.1] (Reference #6, Ch. 5—Deployment Flowchart)

12. a; A *histogram* is a frequency distribution that graphically displays the measurements taken from a process and shows how those data are distributed and centered over a measurement scale. The skewed distribution is unsymmetrical because a natural limit prevents outcomes on one side. The distribution's peak is off center toward the limit and a tail stretches away from it. [I.E.1] (Reference #6, Ch.5—Histogram)

13. d; A *Pareto chart* is a simple method for displaying count or discrete data. A Pareto chart is a bar graph. The length of the bars represents frequency or cost (money or time), and they are arranged in order from longest on the left to shortest on the right. The Pareto chart uses the concept of the 80/20 principle. For example, in a given process, 80% of the defects are caused by 20% of the causes. [I.E.1] (Reference #6, Ch.5—Pareto Chart)

14. a; A *check sheet* is a data collection method for capturing count or discrete data. Check sheets are used to observe or review a process, usually during execution of the process. Check sheets precategorize potential outcomes for data collection using sets of words, tally lists, or graphics. This simple tool provides a method for

easy collection of the data. By collecting data on a check sheet, common patterns or trends can be identified. [I.E.1] (Reference #2, Ch.7—D. Tools)

15. d; The most effective methods of process improvement utilize teams representing process owners and stakeholders. Process owners have responsibility for the execution and implementation of the process and are considered subject matter experts. Stakeholders are those who have a vested interest in the process and/or its products and outputs. Stakeholders include customers, suppliers, employees, investors, and communities. [II.A.3] (Reference #2, Ch.4—A.5)

16. a: The ultimate goal of an effective supply chain is customer loyalty. The systematic nature of a process management (PM) system works well to address the potential weak links in the chain—strategy and design, information flow, and operational control. [II.A.5] (Reference #1, Ch.13—The Process of Supply Chains)

17. d: A supply chain is complex, made up of many suppliers located around the world, each of which has its own supply chain. The purpose of a supply chain is to place products and services timely and correctly in stores or at customer locations. The supply chain must be designed, directed, and managed as a process, not as a series of order and shipping transactions. [II.A.5] (Reference #1, Ch.13—Supplier Relationship)

18. d; An *affinity diagram* is a display of brainstormed ideas, survey results, and any type of input arranged by affinity or any type of overarching theme. The *affinity diagram* complements brainstorming and organizes a large number of ideas into their natural relationships. [II.B.4] (Reference #3, II.B.4)

19. b; The Juran trilogy is based on three managerial processes: *quality planning*, *quality control*, and *quality improvement*. Without change, there will be a constant waste; during change, there will be increased costs. But after the improvement, margins will be higher, and the increased costs get recouped. [IV.C.2] (Reference #2, Ch.1—A.1)

    quality trilogy—A three-stage approach to managing for quality. The three stages are *quality planning* (developing the products and processes required to meet customer needs), *quality control* (meeting product and process goals), and *quality improvement* (achieving unprecedented levels of performance). Attributed to Joseph M. Juran. [IV.C.2] (Reference #1, Appendix C)

20. c; A *control chart* is used to identify special cause and common cause variation. Common cause variation is random variation occurring in the process and presents as random fluctuations within the control limits. Special cause variation is nonrandom variation and presents as nonrandom patterns or points outside the control limits. A *trend* is considered nonrandom and therefore special cause variation. [V.B.2] (Reference #1, Ch.9—Control Chart)

The measures of dispersion—range, variance, and standard deviation—are calculated as follows.

Wk 1	Wk 2	Wk 3	Wk 4	Wk 5	Wk 6	Wk 7	Wk 8	Wk 9	Wk 10
4.0	2.0	3.0	2.0	0.0	0.0	1.0	1.0	2.0	2.0

21. a; To calculate the *range*, subtract the lowest number from the highest number. [III.A] (Reference #3, III.A)

$$\text{Range} = 4.0 - 0.0 = 4.0$$

22. a; To calculate the *variance*, subtract the average from each data point and sum the differences. [III.A] (Reference #3, III.A)

$$s^2 = \frac{\Sigma(x_i - \bar{x})^2}{n-1}$$

s^2 = Variance

x_i = Term in data set

$\bar{x}$ = Sample mean

Σ = Sum

n = Sample

	x	$\bar{x}$	$(x - \bar{x})$	$(x - \bar{x})^2$
X^1	4		2.3	5.29
X^2	2		0.3	0.09
X^3	3		1.3	1.69
X^4	2		0.3	0.09
X^5	0		1.7	2.89
X^6	0		1.7	2.89
X^7	1		0.7	0.49
X^8	1		0.7	0.49
X^9	2		0.3	0.09
X^{10}	2		0.3	0.09
Σ	17	1.7		14.1

$$s^2 = \sqrt{\frac{\Sigma(x_i - \bar{x})^2}{n-1}} = \frac{14.1}{10-1} = 1.57$$

23. d; To calculate the *standard deviation*, take the square root of the variance. [III.A] (Reference #3, III.A)

$$s^2 = \sqrt{\frac{\Sigma\left(x_i - \bar{x}\right)^2}{n-1}} = \frac{14.1}{10-1} = 1.57$$

$$s = \sqrt{\frac{\Sigma\left(x_i - \bar{x}\right)^2}{n-1}} = \sqrt{1.57} = 1.2516$$

24. c; The *operational definition* in a data collection plan is a measurement defined in such a way as to achieve repeatable results from multiple observers. [III.B.1] (Reference #3, III.B.1)

25. b; The three elements of "what to measure" in the data collection plan are:

1. Measure	2. Type of measure	3. Operational definition
Name of parameter or condition to be measured	*X* or *Y* attribute or discrete data, product, or process data	Measurement defined in such a way as to achieve repeatable results from multiple observers

[III.B.1] (Reference #3, III.B.1)

26. c; *Continuous data* are a type of *quantitative data. Continuous data* can be any number of possibilities between two whole numbers; for example, the length of a widget can be 10 mm or 10.1 mm or 10.001 mm, and so on. It can go any number of decimals or fractions that the measurement method and technology will practically allow. Examples of continuous data are product dimensions and service quality delivery cycle times. [III.B.2] (Reference #2, Ch.13—D.1; Reference #3, III.B.2)

27. a; A *ratio* measurement scale has a rational zero point for the scale. For example, money is an example of a ratio measurement scale. Money can be counted as five cents, 10 cents, and so on. If you have zero money, it implies the absence of money. Given that money has a true zero point it makes sense to say that a person with $50.00 has twice as much money as a person who has $25.00. [III.B.2] (Reference #2, Ch.13—D.1)

28. b; The *telephone interview* is a type of survey used when:

    a. Most possible answers to the questions are known.

    b. Data are needed quickly.

    c. A high response rate is needed.

    d. People resources are available for making calls.

    [III.B.3] (Reference #6, Ch.5—Telephone Interview)

29. c; The *face-to-face interview* is a type of survey used when:
    a. The group to be surveyed is small.
    b. Ample resources (time, people, and possibly travel expenses) are available.
    c. Possible answers to the questions are not known, such as when you first begin studying an issue.
    d. The questions to be asked are sensitive.
    e. The people to be surveyed are high-ranking, important, or otherwise deserving of special attention.
    f. Close to 100% response rate is needed.

    [III.B.3] (Reference #6, Ch.5—Face-to-Face Interview)

30. d; The *questionnaire* is a type of survey used when:
    a. Possible responses to questions are known and need only be quantified, such as after an initial set of interviews or focus groups has identified key issues.
    b. Collecting data from a large group.
    c. Data are not needed immediately.
    d. A low response rate can be tolerated.
    e. Resources (money, people) for collecting data are limited.

    [III.B.3] (Reference #6, Ch.5—Questionnaire)

31. b; *Bias* is the difference between absolute value and true value with respect to a standard master at various measurement points of the measuring range. (In practice, accuracy and bias often are used interchangeably.) Understanding bias during the *measure* phase helps process owners understand why the equipment is not accurate and that there may be a need for calibration and adjustment of bias closer to the true value. [III.C.1] (Reference #3, III.C.1)

32. b; *Linearity* is defined as the accuracy of measurement at various points in the measuring range of the equipment. Understanding linearity during the *measure* phase helps process owners understand why measurement is inconsistent across the measurement range of the equipment. [III.C.1] (Reference #3, III.C.1)

33. b; *Reproducibility* is the variation in measurement made by two or more appraisers multiple times. Reviewing reproducibility during the *measure* phase helps process owners understand the variability due to the human inability to reproduce the same measurement trial after trial. [III.C.1] (Reference #3, III.C.1, glossary)

34. a; MSA is a method of checking how well operators are using gages and test equipment and judging whether the measurement system is effective and efficient. The process of testing is sometimes called *gage repeatability and reproducibility* (GR&R). The gage R&R study assesses the gage's accuracy, repeatability, reproducibility, and stability. Results from the study will show whether the variation in the data is due to the measurement system or some other source. [III.C.2] (Reference #4, Tools—Measurement Systems Analysis; Reference #6, Ch.5—Repeatability and Reproducibility Study)

35. a; The biggest difference is that *ANOVA* includes *interactions between operators and samples.* An example of an interaction might be two inexperienced operators obtaining inaccurate measurements at only one end of the measurement range, thus on only three out of 10 samples. [III.C.2] (Reference #6, Ch.5—Repeatability and Reproducibility Study)

36. b; The part-to-operator interaction *p* value is .974. At the 95% confidence level, the rejection region for significance is less than .05. In this case the *p* value is .974, which is greater than .05; therefore, we fail to reject the null hypothesis stating that the part-to-operator interaction is equal to zero. [III.C.2] (Reference #6, Ch.5—Repeatability and Reproducibility Study) (Example data source—Reference #2, Ch.14—III.E)

37. d; *Non-value-added* refers to tasks or activities that can be eliminated with no deterioration in product or service functionality, performance, or quality in the eyes of the customer. The eight types of waste are categorized as non-value-added activities. [IV.A.1] (Reference #1, Ch.7—How to Start the Lean Journey; Reference #3, IV.A.1)

38. c; The purpose of a *failure mode and effects analysis* (FMEA) is to understand the opportunities for failure and the impact of risks in a product or process design, prioritize the risks, and take actions to eliminate or reduce the impact of these risks. [IV.A.2] (Reference #3, IV.A.2)

39. b; The *risk priority number* (RPN) is a numeric assessment of risk assigned to a process, or steps in a process, during an FMEA. It is calculated by multiplying the *severity* rating times the *occurrence* rating times the *detection* rating (S × O × D). [IV.A.2] (Reference #3, IV.A.2)

40. b; *Occurrence* is defined as the likelihood or frequency at which the cause can occur. Typically, the scale is 1 to 10. Higher occurrence is rated at the high end of the scale, lower occurrence at the low end of the scale. [IV.A.2] (Reference #3, IV.A.2)

41. b; The *cause-and-effect* or *X–Y* relational matrix is used to quantify and prioritize the impacts of causes (*X*) on effects (*Y*) through numerical ranking. This relational matrix shows the impact scores, which can be based on objective

measures or subjective ordinal values. [IV.B] (Reference #2, Ch.19—B Cause-and-Effect (X–Y) Relational Matrix)

42. a; A *process* is a set of interrelated or interacting activities that transforms inputs into outputs. A *process flowchart* is a helpful tool during problem investigation. A process is easily understood by visually presenting the process using common flowcharting shapes and symbols. [IV.B] (Reference #3, IV.B)

43. d; The binomial distribution is used to model discrete data and should be applied in situations where each part has just two states, such as:

    a. Good or bad

    b. Accept or reject

    c. Conformance or nonconformance

    d. Success or failure

    Specific conditions of the binomial distribution include (1) the population denoted by N is greater than 50, and (2) the sample size is less than 10% of the population. [IV.C.1] (Reference #3, IV.E.1)

44. b; The mathematical formula for the normal distribution is

$$P(x) = \frac{e^{\frac{(x-\mu)^2}{2\sigma^2}}}{\sigma\sqrt{2\pi}}$$

    [IV.C.1] (Reference #3, IV.E.1)

45. a; The *central limit theorem* suggest that the distribution of averages tends to be normal even when the distribution from which the average data are computed is from nonnormal distributions. Mathematically, if a random variable X has a mean μ and variance σ^2, as the sample size n increases, the sample mean $\overline{X}$ approaches a normal distribution with mean μ and variance σ_x^2. The central limit theorem consists of three statements:

    1. The mean of the sampling distribution of means is equal to the mean of the population from which the samples were drawn.

    2. The variance of the sampling distribution of means is equal to the variance of the population from which the samples were drawn divided by the size of the samples.

    3. If the original population is distributed normally (that is, bell-shaped), the sampling distribution of means will also be normal.

    [IV.E.1] (Reference #3, IV.E.1)

46. d; *Special cause* variation is defined as variation that arises because of special circumstances. They are not an inherent part of a process. Special causes are also referred to as *assignable causes*. Special cause variation can be identified as points outside the control limits or through patterns that suggest nonrandom variation. [IV.C.2] (Reference #3, IV.E.2, glossary)

47. c; *Common cause* variation is variation that is inherent in a process over time. The variation affects every outcome of the process and everyone working in the process. Common cause variation displays as random data points within the control limits. [IV.E.2] (Reference #3, IV.E.2)

48. b; *Common causes of variation* are those that are inherent in the process and generally are not controllable by process operators. Common cause variation is also known as *natural variation* and refers to the many sources of variation within a process. Common causes reside in processes within statistical control and can be characterized by *location (process average), spread (piece-to-piece variability),* and *shape (distribution)* for predictability. [IV.E.2] (Reference #3, IV.C.2)

49. a; *Correlation* is finding a relationship between two or more sets of data. It measures the *strength* and *direction* of the relationship between variables. In order to find a correlation, one needs an independent variable (x) that causes an observed variation, dependent variable (y), that is an effect of (x). [IV.F.1] (Reference #3, IV.F.1)

50. b; The *correlation coefficient* r provides both the strength and the direction of the relationship between the independent and dependent variables. Values of r range between –1.0 and +1.0. When r is positive, the relationship between x and y is positive, and when r is negative, the relationship is negative. A correlation coefficient close to zero is evidence that there is no relationship between x and y. The strength of the relationship between x and y is measured by how close the correlation coefficient is to +1.0 or –1.0. [IV.D.1] (Reference #3, IV.F.1)

51. a; The strength of the relationship between x and y is measured by how close the correlation coefficient is to +1.0 or –1.0. The correlation coefficient of +.90 is very close to +1.0 and therefore shows a strong positive correlation. [IV.D.1] (Reference #3, IV—F.1)

52. c; The *least squares method* is a mathematical procedure for identifying the linear equation that best fits a set of ordered pairs by finding values for a (the y-intercept) and b (the slope). The goal of this method is to minimize the total squared error between the values of y and $\hat{y}$. If we denote the predicted value of y obtained from the fitted line as $\hat{y}$, the predicted equation is

$$\hat{y} = \hat{a} + \hat{b}x$$

where $\hat{a}$ and $\hat{b}$ represent estimates of true a and b. Since we need to choose the best-fitting line, we need to define what we mean by "best." For the purpose of getting the best-fitting criteria, the principle of least squares is employed; that is, one has to choose the best-fitting line, the line that minimizes the sum of squares of the deviations of the observed values of y from those predicted. [IV.D.2] (Reference #3, IV.F.2)

53. b; In simple linear regression, where the linear equation $\hat{y} = a + bx$, b represents the slope of the line. [IV.D.2] (Reference #3, IV.F.2)

54. d; The *coefficient of determination*, r^2, is a number between 0 and 1 that measures how well the data fit the line. If $r^2 = 1$, the line fits the data perfectly. As r^2 gets smaller, the line's fit becomes poorer, and predictions made from it will be less accurate. The coefficient of determination r^2 is the proportion of y's variation that is explained by the regression line. Because most data points don't fall exactly on the line, the rest of the proportion $(1 - r^2)$ is error. [IV.F.2] (Reference #6, Regression Analysis)

55. b; A *type I error* results when the null hypothesis is rejected when it is actually true. For example, incoming products are good but were labeled defective. This type of error is also called α (alpha) and referred to as the *producer's risk* (for sampling). [IV.E] (Reference #3, IV.E)

56. a; The *power* of a statistical test in sampling is the probability of correctly rejecting a false null hypothesis (type II error). A power of 0.8 and above is typically required for making a conclusion. A power of 0.9 or more may be required in some situations based on the risk to the organization. A power of 0.8 means an experiment with the current sample size has an 80% likelihood of identifying a significant difference (more than 1% defectives) when one truly exists, and a 20% likelihood it will incorrectly identify a significant difference when the difference does not exist. [IV.E] (Reference #3, IV.E; Reference #2, Ch.17—B.1)

57. b; *Kaizen (continuous improvement)* is a Japanese term for change for improvement, or improving processes through small, incremental steps. Many people refer to this gradual change as *continual improvement.* [V.A.1] (Reference #2, Ch.2—B.1)

58. a; *Kaizen blitz/events* are used to provide quicker implementation results. Kaizen blitz/events are conducted by assembling a cross-functional team for three to five days and reviewing all possible options for improvement in a breakthrough effort. In contrast, *kaizen (continuous improvement)* is a Japanese term for change for improvement, or improving processes through small, incremental steps. [V.A.1] (Reference #2, Ch.2—B.1)

59. a; Masaaki Imai made popular the practice of *kaizen*, a strategy for making improvements in quality in all business areas. *Kaizen* focuses on implementing small, gradual changes over a long time period. When the strategy is fully utilized, everyone in the organization participates. *Kaizen* is driven by a basic belief that when quality becomes ingrained in the organization's culture and people, the quality of products and services will follow. [V.A.1] (Reference #1, Ch.7—Kaizen)

60. c; *PDSA* is a process intended to be a guiding light for the way we work and do our jobs. Building on the work of Walter Shewhart in the 1930s, *Dr. W. Edwards Deming* changed the improvement model PDCA (plan, do, check, act) to *PDSA* (plan, do, study, act). The latter model is the foundation for various processes such as advanced quality planning, problem solving, Six Sigma, and process improvement. [V.A.2] (Reference #4, Tools—PDSA)

61. c; The *PDCA cycle* is sometimes referred to as the *Shewhart cycle*. PDCA evolved from 1939 to 1994, from Walter A. Shewhart to W. Edwards Deming. It is a four-step process for quality improvement. In the first step (plan), a plan to effect improvement is developed. In the second step (do), the plan is carried out, preferably on a small scale. In the third step (check), the effects of the plan are observed. In the last step (act), the results are studied to determine what was learned and what can be predicted. [V.A.2] (Reference #3, V.A.2)

62. b; A *cost/benefit analysis* involves calculating both the applicable measurable costs that will be incurred and the benefits of the costs avoided. Financial measures for cost/benefit analysis include (1) return on investment, (2) payback period, (3) net present value, and (4) internal rate of return. [V.A.3] (Reference #3, V.A.3)

63. a; The *payback period* is the time it takes to recoup the investment made in a project; the shorter the period, the better. [V.A.3] (Reference #3, V.A.3)

64. c; The *payback period* is the time it takes to recoup the investment made in a project; the shorter the period, the better. [V.A.3] (Reference #3, V.A.3)

65. d; A *control plan* is a document summarizing all the key information that an operator needs to know to manage, control, and monitor a process. ASQ defines a control plan as "written descriptions of the systems for controlling part and process quality by addressing the key characteristics and engineering requirements." The control plan (updated as needed and kept current) explains how to control the workflow in your process. [V.B.1] (Reference #3, V.B.1)

66. b; An $\bar{X}$ *and* R *control chart* is constructed using a rational subgroup. A *rational subgroup* is a sample set that is sufficient to determine common cause scenarios. Normally, the average of a subgroup is used. In rational subgrouping (1) the division of observations into rational subgroups is key, (2) success of control charting depends on the selection of subgroups (size and frequency of sampling),

and (3) selection should result in groups as homogeneous as possible. [V.B.2] (Reference #3, V.B.2)

67. d; A *rational subgroup* is a sample set that is sufficient to determine common cause scenarios. Normally, the average of a subgroup is used. In rational subgrouping (1) the division of observations into rational subgroups is key, (2) success of control charting depends on the selection of subgroups (size and frequency of sampling), and (3) selection should result in groups as homogeneous as possible. [V.B.2] (Reference #3, V.B.2)

68. d; A *run chart* is a line graph showing a process measurement on the vertical axis and time on the horizontal axis. Often, a reference line shows the average of the data. The run chart reveals patterns in the data over time. Unlike a control chart, it does not show control limits. [V.B.2] (Reference #6, Ch.5—Run Chart)

69. b; A *work instruction* is a step-by-step description of how to complete a task. Work instructions (WI) are detailed instructions of how a process is carried out step-by-step. There may be multiple WI under a given SOP depending on the type and complexity. [V.B.3] (Reference #3, V.B.3)

70. b: Standard operating procedures (SOPs) are known by many names: work instructions, level-three ISO 9001 documentation, operating guides, job aids, standard job practices, and more. [V.B.4] (Reference #4, Tools—Standard Operating Procedure)

71. c; Six Sigma is a term describing the overall concept of *continual improvement.* [I.A] (Reference #4, What Is Six Sigma?)

72. b; Statistically speaking, *sigma* is a term indicating to what extent a process varies from perfection. The quantity of units processed divided into the number of defects actually occurring, multiplied by one million, results in *defects per million.* [I.A] (Reference #1, Ch.7—DMAIC Methodology)

73. d; After Motorola started promoting their Six Sigma methodology in the late 1980s, there have been many skeptical of its true value. Even Jack Welch of General Electric (GE) initially dismissed the idea of Six Sigma as a passing fad in the early 1990s. However, once GE had a successful launch in one of its divisions, Six Sigma quickly became a driving force that started spreading across various industries in the mid to late 1990s. [I.A] (Reference #2, Ch.1—A.1)

74. a; If used correctly, Six Sigma will:

    1. Create an infrastructure for managing improvement efforts and focus resources on those efforts

    2. Ensure that those improvement efforts are aligned with client and stakeholder needs

3. Develop a measurement system to monitor the impact of improvement efforts

[I.A] (Reference #1, Ch.7—Six Sigma Concepts and Tools)

75. b; A *value stream* is the series of activities that an organization performs, such as order, design, produce, and deliver products and services. A value stream often starts with a supplier's supplier and ends at the customer's customer. Wastes are both explicit and hidden along a value stream. [I.B] (Reference #3, I.B)

76. a; This concept is visually illustrated with a lean tool called the *value stream map*. This map uses simple graphics and icons to illustrate the movement of material, information, inventory, work-in-progress, operators, and so on. [I.B] (Reference #3, I.B)

77. c; A definition of *lean*, used by the Manufacturing Extension Partnership (of NIST/MEP, part of the U.S. Department of Commerce), is "a systematic approach in identifying and eliminating waste (non-value-added activities) through continuous improvement by flowing the product at the pull of the customer in pursuit of perfection." Lean focuses on value-added expenditure of resources from the customers' viewpoint. [I.B] (Reference #1, Ch.7—Lean Concepts and Tools)

78. a; *Just-in-time* (JIT). JIT is a production strategy promoted by Toyota, and now applied to many organizations, that strives to improve business return on investment by reducing in-process inventory and associated carrying costs. [I.B] (Reference #2, Ch.2—B)

79. d; *Black Belt,* individuals trained in Six Sigma methodologies, statistical tools, basic financial tools, change management, risk assessment, and project management, and well experienced in managing Black Belt projects. [I.C] (Reference #3, I.C t)

80. b; *Master Black Belt*:

    a. Expert on Six Sigma tools and concepts

    b. Trains Black Belts and ensures proper application of methodology and tools

    c. Coaches and mentors Black and Green Belts

    d. Works high-level projects and those that impact multiple divisions or business units

    e. Assists champions and process owners with project selection, management, and Six Sigma administration

    [I.C] (Reference #1, Ch.7—DMAIC Methodology t)

81. d; *Champion*, typically an upper-level manager:

    a. Liaises with senior management

    b. Allocates resources for projects

    c. Determines project selection criteria

    d. Removes barriers hindering the success of the project

    e. Approves completed projects

    f. Implements change

    [I.C] (Reference #3, I.C t)

82. a; *Yellow Belt*:

    a. Willing to commit to the purpose of the team

    b. Understands lean and Six Sigma tools and concepts

    c. Able to express ideas, opinions, suggestions in a nonthreatening manner

    d. Capable of listening attentively to other team members

    e. Receptive to new ideas and suggestions

    f. Able to engage in analysis of Lean Six Sigma tools and concepts

    g. Even-tempered, able to handle stress and cope with problems openly

    h. Competent in one or more fields of expertise needed by the team

    i. Favorable performance record

    j. Willing to function as a team member and forfeit "star" status

    [I.C] (Reference #1, Ch.7—DMAIC Methodology t)

83. b; A team usually comprises *five to nine members* (seven is considered an ideal size) with complementary skills necessary to achieve the goals and objectives of the team. Team composition should be driven by the size and scope of the project; it is possible to have a team of one or two for a smaller project and a large team with sub-teams for a big project. [I.D.1] (Reference #2, Ch.9—F)

84. c; *Virtual teams* enable people from all over the globe to meet via teleconferences, videoconferences, and internet tools such as shared computers. There are many benefits to virtual teaming, the most prevalent being reduced costs and real-time data sharing and updating. [I.D.1] (Reference #2, Ch.9—F)

85. d; *Self-directed teams* tend to share leadership equally among the group members; thus, rather than one person being the leader, they are all responsible for the output of the team. [I.D.1] (Reference #4, Tools—Self-Directed Teams)

86. b; Product development, continual improvement, and problem-solving projects require a *cross-functional team* to be assembled. Representatives from all the different functions will be required in order to obtain the knowledge and experience needed for a project. [I.D.1] (Reference #3, Ch.4—D.1)

87. c; Stage 3: *Norming.* During this stage, members reconcile competing loyalties and responsibilities. They accept the team, the team ground rules or "norms," their roles in the team, and the individuality of fellow members. [I.D.2] (Reference #5, p. 6–7)

88. d; Stage 4: *Performing.* As team members become more comfortable with each other, and as they better understand the work and what is expected of them, they become a more effective unit with everyone working in concert. [I.D.2] (Reference #5, p. 6–8)

89. b; Stage 5: *Adjourning.* Other authors have added adjourning to Tuckman's original model. This stage is the process of closure that occurs when a team has accomplished its mission. The actions include reviewing lessons learned, assessing the achievement of the outputs and outcomes intended, completing documentation, recognizing the team's efforts and celebrating, and formally disbanding the team. This stage is often skipped or inadequately addressed in the team's haste to disband and move on. [I.D.2] (Reference #1, Ch.6—Team Stages)

90. a; The *forming* stage includes these behaviors:

    a. Attempting to define the task and deciding how it will be accomplished

    b. Deciding what information is needed

    c. Determining acceptable team behavior and how to deal with team problems

    d. Establishing team ground rules

    e. Waiting to be told what to do and directing most communication to the team leader

    Other distractors are behaviors observed in subsequent stages of team development. [I.D.2] (Reference #5, p. 6–4)

91. b; *Multivoting* narrows a large list of possibilities to a smaller list of the top priorities or to a final selection. Multivoting is preferable to straight voting

Section 2 Solutions

because it allows an item that is favored by all, but not the top choice of any, to rise to the top. [I.D.3] (Reference #6, Ch.5—Multivoting)

92. c; The *nominal group technique* (NGT) is a structured process that identifies and ranks major problems or issues that need addressing. It can be used to identify the major strengths of a department, unit, or institution or to make decisions by consensus when selecting problem solutions in a business. [I.D.3] (Reference #1, Ch.6—Team Decision-Making Tools)

93. d; *Brainstorming* is a process where a team develops as many ideas concerning a topic as possible, using various creative methods. Brainstorming is a powerful technique for soliciting ideas, and it is used extensively at every stage of improvement or problem solving. This tool intentionally encourages divergent thinking through which, hopefully, all possible causes are identified. This is a team exercise and requires a good facilitator to get the ideas flowing. Brainstorming has two phases: the *creative* phase, in which a large number of ideas are generated, and the *evaluation* phase, in which the ideas generated are looked at for usefulness or applicability. The two phases should be separated by a time break, as different parts of the brain are used in each phase. At a minimum, a 10-minute break should be taken after the creative phase versus going directly to the evaluation phase. [I.D.3] (Reference #3, Ch.4—D.3)

94. b; *Brainwriting* is a nonverbal form of brainstorming. Team members write down their ideas individually. Ideas are shared through an exchange of papers, then additional ideas are written down.

    When to use: Try brainwriting instead of brainstorming for any of the following reasons:

    a. When a topic is too controversial or emotionally charged for a verbal brainstorming session

    b. When participants might feel safer contributing ideas anonymously

    c. To encourage equal participation, typically when verbal brainstorming sessions are dominated by a few members

    d. When some group members think better in silence

    e. When ideas are likely to be complex and require detailed explanation

    [I.D.3] (Reference #6, Ch.5—Brainwriting)

95. c; Minutes. The meeting minutes are a record of the meeting. Minutes are essential to ensure that key decisions made and the actions agreed on by the team members are formally recorded and to keep the team members accountable. The minutes should be well drafted and unambiguous, and they should indicate the date and time of the completed meeting, meeting host, attendees, topics

covered, decisions made, actions assigned (what, who, when), minutes reviewed and approved, and scribe name. One might also include "parking lot" items that are pending detailed discussions. Meeting minutes are important to project continuity. [I.D.4] (Reference #3, Ch.4—D.4)

96. d; An agenda should be set for these reasons:

    a. It is an essential step in determining whether a meeting is needed and, if it is, in building a successful one.

    b. It spells out *how* the outcomes can be accomplished (process).

    c. An agenda establishes a *time frame*.

    d. It helps identify *who* needs to be present. [I.D.4] (Reference #4, Tools—Team Meetings)

97. a; Take time periodically throughout the project to stop and ask, "How are we doing?" This can be one of the most important and difficult activities a team can undertake. Self-critique can help a team identify problems early, before they become crises. We recommend that teams pause to review their work at the end of each major step in their work plans. [I.D.4] (Reference #5, p. 3–60)

98. b; *Logistics* is defined as a process involving planning, implementing, and controlling an efficient, cost-effective flow and storage of raw materials, in-process inventory, finished goods, and related information from point of origin to point of consumption for the purpose of conforming to customer requirements. Breakdowns in the planning and implementing phases can substantially and negatively impact the work of a team. For example, consider some of the issues for team members regarding meetings:

    a. If team members are not in the same building, what connectivity and communication problems can arise?

    b. The logistics of getting people together for kaizen events can be frustrating (who carries on the work when the member is at the event?).

    c. Selecting conference rooms, technology, room setup, and so on creates logistic nightmares (who coordinates these arrangements?).

    d. Agendas, notes, action items, and so on are part of the logistics of team meetings (who attends to these functions?).

    [I.D.4] (Reference #1, Ch.6—Other Barriers)

99. c; A *scatter diagram* is a chart in which one variable is plotted against another to determine whether there is a correlation between the two variables. These diagrams are used in plotting the distribution of information in two dimensions.

Scatter diagrams are useful in rapidly screening for a relationship between two variables.

A scatter diagram shows the pattern of relationship between two variables that are thought to be related. For example, is there a relationship between outside temperature and cases of the common cold? As temperatures drop, do colds increase? The more closely the points hug a diagonal line, the more likely it is that there is a one-to-one relationship.

The purpose of the scatter diagram is to display what happens to one variable when another variable is changed. The diagram is used to test a theory that the two variables are related. The slope of the diagram indicates the type of relationship that exists. [I.E.1] (Reference #1, Ch.9—Scatter Diagram)

100. d; A *check sheet* is a structured, prepared form for collecting and analyzing data. This is a generic tool that can be adapted for a wide variety of purposes.

     When to use:

     a. When data can be observed and collected repeatedly by the same person or at the same location

     b. When collecting data on the frequency or patterns of events, problems, defects, defect location, defect causes, and so forth

     c. When collecting data from a production process

     [I.E.1] (Reference #6, Ch.5—Check sheet)

101. a; When using the cause-and-effect diagram, it is best to try to keep an open mind and to work as a team to view and discuss what the system or process is doing. You want to capture everything you can about the process, looking for the real state of the system, not just what you think is happening. Besides using the five W's and one H in creating the cause-and-effect diagram, most people start with the *6 M's*:

     1. Man (people—operator)
     2. Machine (equipment)
     3. Methods (operating procedures)
     4. Materials
     5. Measurement
     6. Mother nature (environment)

     [I.E.1] (Reference #4, Tools—Cause and Effect Diagram)

102. b; The *run chart* is used to identify patterns in process data. All of the individual observations are plotted in a time sequence, and a horizontal reference line is drawn at the median. A run chart is typically used when the subgroup size is one. When the subgroup size is greater than one, the subgroup means or medians are calculated and connected with a line, similarly to a control chart. However, run charts are different from control charts (for example, $\bar{X}$ and R charts); *run charts do not have statistical control limits to monitor variation.* There are also related statistical tests that can be performed to detect any nonrandom behavior. [I.E.1] (Reference #3, Ch.5—E.1)

103. c; A *cost-of-poor-quality analysis* is a way of studying a process's flowchart to identify potential problems. *Cost of poor quality* means costs incurred because things are not done right the first time, every time. The analysis helps a team look critically at individual steps of a process to find opportunities for improvement.

When to use:

a. When flowcharting a process, to be sure that *cost-of-poor-quality* activities are included

b. After flowcharting a process, to identify problems, potential causes, and areas in which to concentrate improvement efforts.

[I.E.2] (Reference #6, Ch.5—Cost-of-Poor-Quality Analysis)

104. d; *Cost of quality* is a methodology that allows an organization to determine the extent to which organizational resources are used for activities that *prevent poor quality,* that appraise the quality of the organization's products or services, and that result from internal and external failures. Having such information allows an organization to determine the potential savings to be gained by implementing process improvements.

Quality-related activities that incur costs may be divided into *prevention costs,* appraisal costs, and internal and external failure costs.

Prevention costs are costs incurred to prevent or avoid quality problems. These costs are associated with the design, implementation, and maintenance of the quality management system. They are planned and incurred before actual operation, and they could include:

- Product or service requirements—establishment of specifications for incoming materials, processes, finished products, and services
- Quality planning—creation of plans for quality, reliability, operations, production, and inspection

- Quality assurance—creation and maintenance of the quality system
- Training—development, preparation, and maintenance of programs.

[I.E.2] (Reference #1, Ch.8—Cost of Poor Quality or Cost of Quality)

105. a; *DPU* is calculated as the total number of defects divided by the total number of products produced in some time period (for example, per day). This metric is used when the area under inspection is large in size or volume, for example, the number of defects on a painted chassis or textile. [I.E.2] (Reference #3, Ch.5—E.2)

106. b; *Cycle time* is the elapsed time that it takes to complete a process from beginning to end. *Cycle time reduction* comprises the action(s) taken to reduce the overall process time from start to finish. [I.E.2] (Reference #1, Ch.7—Cycle Time Reduction)

107. c; Methods used to *identify customers* include:
    - Brainstorming
    - SIPOC
    - Marketing analysis data
    - Tracking a product or service to delivery.

    [II.A.1] (Reference #2, Ch.5—B.1)

108. d; The following figure is an example from a consolidated call center supporting a large county health department:

Translation of VOC to customer requirements to metrics

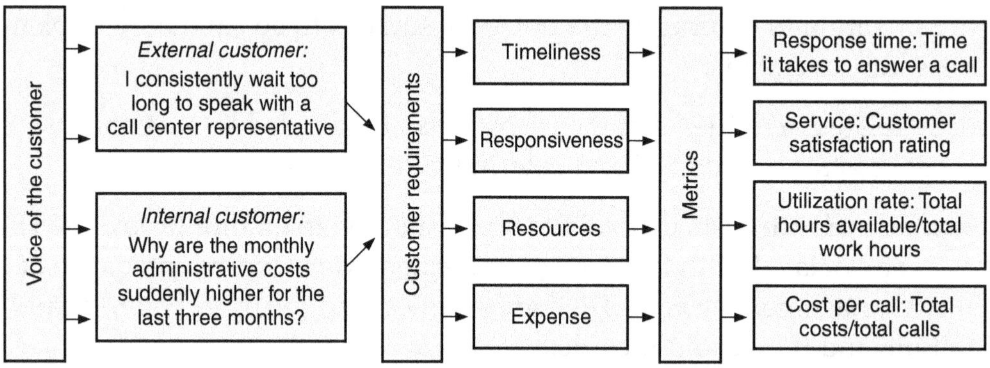

Source: Reproduced by permission from R. Bialek, G. L. Duffy, and J. W. Moran, *The Public Health Quality Improvement Handbook* (Milwaukee: ASQ Quality Press, 2009), 227.

The department's leadership met with a broad base of community stakeholders, clients, and service partners to learn of their priority needs for using the call center. Prompt response time was a major external customer requirement, identified in the statement "I consistently wait too long to speak with a call center representative." [IIA.1] (Reference #1, Ch.7—DMAIC Methodology)

voice of the customer—An organization's efforts to understand the customers' needs and expectations ("voice") and to provide products and services that truly meet such needs and expectations.

[II.A.1] (Reference #1, glossary)

109. b; *Stakeholder analysis* identifies individuals or groups with an interest in an issue. Based on assessments of the interest, influence, and importance of each stakeholder, actions are planned to change those assessments or to work within them to ensure success of a project or plan. [II.A.1] (Reference #6, Tools—Stakeholder Analysis)

110. c; Translating customer needs into quantifiable *critical-to-quality* (CTQ) characteristics is the execution side of the business. It involves taking customer inputs and designing the product or service while ensuring that the product is manufacturable, and the service is consistently delivered to the customer.

    Most organizations use a disciplined design and development process that encompasses many tools like the Pugh concept selection matrix, failure mode and effects analysis (FMEA), and the process capability study. These tools, along with a structured approach to product development, help ensure that customer needs are translated into *CTQ requirements* for the products and services. [II.A.1] (Reference #3, Ch.6—A.1)

111. a; It is common to require that project proposals include precise statements of the problem definition and some preliminary measures of the seriousness of the problem, including its impact on the goals of the organization. For some managers, these will be the criteria that define which projects to start first based on *which ones save the most bottom-line dollars* ($\bar{\bar{S}}$). [II.A.2] (Reference #2, Ch.4—A.1)

112. d; In some companies the project selection group assigns some projects to Six Sigma teams and other projects to teams using other methodologies. For example, *problems involving extensive data analysis and improvements using designed experiments would likely be assigned to a Six Sigma team,* whereas a process improvement not involving these techniques might be assigned to a lean manufacturing team employing kaizen tools. [II.A.2] (Reference #2, Ch.4—A.1)

113. b; A project selection group, which includes *Master Black Belts,* Black Belts, organizational champions, and key executive supporters, establishes a set of criteria for project selection and team assignments. [II.A.2] (Reference #3, Ch.6—A.2)

114. d; The DMAIC methodology. Improvement teams use the DMAIC methodology to root out and eliminate the causes of defects through the following planning and implementation phases:

 D—*Define* a problem or improvement opportunity

 M—*Measure* the existing process performance

 A—*Analyze* the process to determine the root causes of poor performance; determine whether the process can be improved or should be redesigned

 I—*Improve* the process by attacking root causes

 C—*Control* the improved process to hold the gains

 In the *define phase,* the defect and the *scope* of the effort are determined. Project champions typically partner with a Master Black Belt to develop the intended outcome and criteria under which the Black Belt will operate. [II.A.2] (Reference #1, Ch.7—The DMAIC Methodology)

115. c; *Stakeholders* are those who have a vested interest in the process and/or its products and outputs. Generally, stakeholders of an organization include customers, suppliers, employees, investors, and communities. Stakeholder interest in and involvement with the process may change over time depending on economic, contractual, and other influences. [II.A.3] (Reference #3, Ch.6—A.3)

116. b; *Subject matter experts* (SMEs) are those who have demonstrated skill and competency in an area that is important to the existence and sustainability of the business. [II.A.3] (Reference #3, Ch.6—A.3)

117. a; *End users* are external customers who purchase products/services for their own use. [II.A.3] (Reference #1, glossary)

118. c; *Process owners* are those responsible for the definition, execution, maintenance, and improvement of the process; in some cases, process owners may also be referred to as *subject matter experts.* Personnel involved with process design usually have a specific interest in systems, subprocesses, and individual steps within processes. [II.A.3] (Reference #2, Ch.4—A.5)

119. d; When identifying the process, it is important to recognize that processes usually affect multiple departments and organizations. Crossing functional areas (departments) and organizations (suppliers, intercompany) can add challenges to an improvement project. The first step in recognizing the challenges is to understand the organizations and functional areas involved with the process. As noted, the *SIPOC diagram* can help in identifying these organizations and functional areas as process suppliers and customers. [II.A.4] (Reference #3, Ch.6—A.4)

120. a; SIPOC stands for *suppliers–inputs–process–outputs–customers.* A *SIPOC diagram* shows a high-level flowchart of a process and lists all suppliers, inputs, outputs, and customers. A SIPOC diagram provides a quick, broad view of key elements of a process.

When to use:

a. At the beginning of a project, to help define the important elements of the project

b. When it is not clear what the process inputs are, who supplies them, what the outputs are, or who the customers are

c. When there are many suppliers, inputs, outputs, and/or customers

[II.A.4] (Reference #6, Tools—SIPOC Diagram)

121. c; *SIPOC analysis.* Process improvement efforts are often focused on rectifying a situation that has developed in which a process is not operating at its normal level. However, much of continual improvement involves analyzing a process that may be performing as expected, but where a higher level of performance is desired. A fundamental step in improving a process is to understand how it functions from a process management perspective. This can be understood through an analysis of the process to identify the suppliers–inputs–process–outputs–customers (SIPOC) linkages. It begins with defining the process of interest. *The outputs that the process creates that go to customers are listed on the right side.* Suppliers and what they provide to enable the process (the inputs) are similarly shown on the left side. Once this fundamental process diagram is developed, two additional items can be discussed: (1) measures that can be used to evaluate performance of the inputs and outputs, and (2) the information and methods necessary to control the process. [II.A.4] (Reference #1, Ch.1—Taking a Systems View of Improvement)

122. d; When planning to study a process or system, it is very important to first identify the *boundaries* to work within. There are many ways to do this; most are not complex and can be easily implemented with common process knowledge and some investigation.

When defining a process, it is important to define its *start and end points—its boundaries.* If, for example, a team is charged with improving a process, team members need to know these process boundaries. Cross-functional processes may incur subprocess boundaries defined by the organizational structure, geography, and so on. [II.A.4] (Reference #3, Ch.6—A.4)

123. a; A *charter* is a document stating the purposes of the project. It serves as an informal contract that helps the team stay on track with the goals of the organization. Each charter should contain the following points:

- *Problem statement.* This is a statement of what needs to be improved.
- *Purpose.* Establishes goals and objectives of the team.
- *Benefits.* States how the enterprise will fare better when the project reaches its goals.
- *Scope.* Provides project limitations in terms of budget, time, and other resources.
- *Results.* Defines the criteria and metrics for project success—including the baseline measures and improvement expectations.

[II.B.1] (Reference #2, Ch.6—C.1)

124. b; *goal*—A statement of general intent, aim, or desire; it is the point toward which management directs its efforts and resources; goals are usually nonquantitative and are measured by supporting objectives. [II.B.1] (Reference #1, glossary)

125. c; To determine *project goals, project baseline data* are collected. The data can be related to items where improvement is required. Progress is made in comparison with the baseline data. For example:

    Eight nonfunctional assemblies manufactured from March 10 to March 20, 2015, were returned by a customer in Fremont, California. The goal is to have *zero* customer returns.

    Twenty percent of the calls made by customers from Alameda County in California in 2014 indicate the average permit approval time was two weeks against the specified one week. The goal is to have 0% complaints.

    [II.B.1] (Reference #3, Ch. 7—B.1)

126. d; Understand your project: one of the team's first tasks is to understand the boundaries of its project. *Review the charter.* Explain the need for improvement, why it's important, timelines, schedules, and the project's boundaries. Outline available resources such as budget, time, people, and so on. Discuss what expertise or technical abilities you might need that aren't represented by team members. Explain what access the team will have to outside experts and technology. On a more detailed level, explain what access the team will have to typing or copying services. [II.B.1] (Reference #5, p. 3–48).

127. a; *Communication planning worksheet* instructions: *Identify some of the people who will be involved in or affected by your project.* Ask them what their main concerns are and make some preliminary decisions about how you will keep them informed throughout the project. Incorporate actions to address their concerns into your work plan. [II.B.2] (Reference #5, p. 1–13)

128. b; Communication is key to the outcome of a project, be it a simple or complex project. A simple project may require daily or weekly meetings with project members meeting face-to-face or through virtual media to update the status of the project and discuss any new risks encountered.

A complex project, however, may require *a formal communication plan*. This means setting up a communication protocol (what, who, where, when, how, and how frequently), escalation protocol, escalation threshold (when to escalate), communication effectiveness verification, and so on. Adding to the challenge is a project team that is spread across geographies (distance and time zones), has different cultures, and has a varying level of infrastructure. [II.B.2] (Reference #3, Ch.7—B.2)

129. a; Too often, operators use the tools and processes of Six Sigma as independent events to satisfy some issue of the moment. Using project planning, you should look at the entire system and strive for synergy as you use *communication* and the tools and processes to contribute to continual improvement. Project planning becomes the tracking system to ensure that all the elements, tools, processes, communications, and so forth are brought together as a whole system for doing the work that you do in the workplace.

Effective project planning requires skills in the following areas:

- Information processing
- *Communication*
- Resource negotiations
- Securing commitments
- Incremental and modular planning
- Assuring measurable milestones
- Facilitating top management involvement

At times, operators may become part of project plans that are developed. Working with such plans should help ensure the successful demonstration of continual improvement and the satisfaction of customers' wants and needs. [II.B.3] (Reference #4, Tools—Project Planning)

130. b; *Work breakdown structure* (WBS) is a project management planning tool by which a project is decomposed into tasks, subtasks, and units of work to be performed, and displayed as a tree-type chart. [II.B.3] (Reference #1, glossary)

131. c; Activities identified from start to finish are called the *work breakdown structure* (WBS). WBS is developed by the project team members by brainstorming. First, high-level activities are identified, and then the team drills down to the sub-activities for every high-level activity. By identifying these activities, the project manager can accurately know the progress and delays and can assign or reallocate resources to ensure that the project stays on schedule. [II.B.3] (Reference #3, Ch.7—B.3)

132. d; A *Gantt chart* is a bar chart that shows the tasks of a project, when each must take place, and how long each will take. As the project progresses, bars are shaded to show which tasks have been completed. People assigned to each task also can be shown.

When to use:

a. When scheduling and monitoring tasks within a project

b. When communicating plans or status of a project

c. When the steps of the project or process, their sequence, and their duration are known

d. When it is not necessary to show which tasks depend on completion of previous tasks

[II.B.3] (Reference #6, Tools—Gantt Chart)

133. a; *Matrix diagram. A matrix diagram is typically used to discover and illustrate relationships between two groups of items.* In the figure below, the two groups are the units of a training course and the objectives of the course. The items in one group are listed across the top of the chart, and the items in the other group are listed down one side.

		Unit								
		1	2	3	4	5	6	7	8	9
Objectives	Review basics	⊙		⊙	○			Δ		
	Math skills	○	Δ	⊙			⊙	○		
	Communication skills						○		Δ	
	Attitude/motivation								Δ	
	Sketching			⊙			○			
	Ohm's law				⊙	○	Δ			
	Kirkoff's law					⊙	○			
	Thevinev's law						Δ		○	
	Heisenberg's uncertainty principle									

⊙ = Strong relationship ○ = Moderate relationship Δ = Weak relationship

Conclusions: Thevinev's law and communication skills covered only weakly
Attitude/motivation barely covered
Heisenberg's uncertainty principle not covered
Unit 2 contributes very little toward objectives
Unit 9 contributes nothing toward objectives

The team examines each square in the matrix and enters one of three symbols or leaves it blank, depending on the relationship between the items in the row and column represented by the square. The most conventional symbols are shown in the example although letters and numbers are sometimes used. The team then examines the completed matrix and discusses possible conclusions. [II.B.4] (Reference #2, Ch.7—D)

134. b; *Relations diagram (interrelationship digraph).* The *relations diagramming* method is a technique developed to clarify intertwined causal relationships in a complex situation to find an appropriate solution. Relations diagrams can be used to:

a. Determine and develop quality assurance policies

b. Establish promotional plans for total quality control introduction

c. Design steps to counter market complaints

d. Improve quality in the manufacturing process (especially in planning to eliminate latent defects)

e. Promote quality control in purchased or ordered items

f. Provide measures against troubles related to payment and process control

g. Promote small group activities effectively

h. Reform administrative and business departments

The digraph in the figure below shows some of the interrelating factors pertaining to ongoing and proposed projects. [II.B.4] (Reference #1, Ch.9—Relations Diagram)

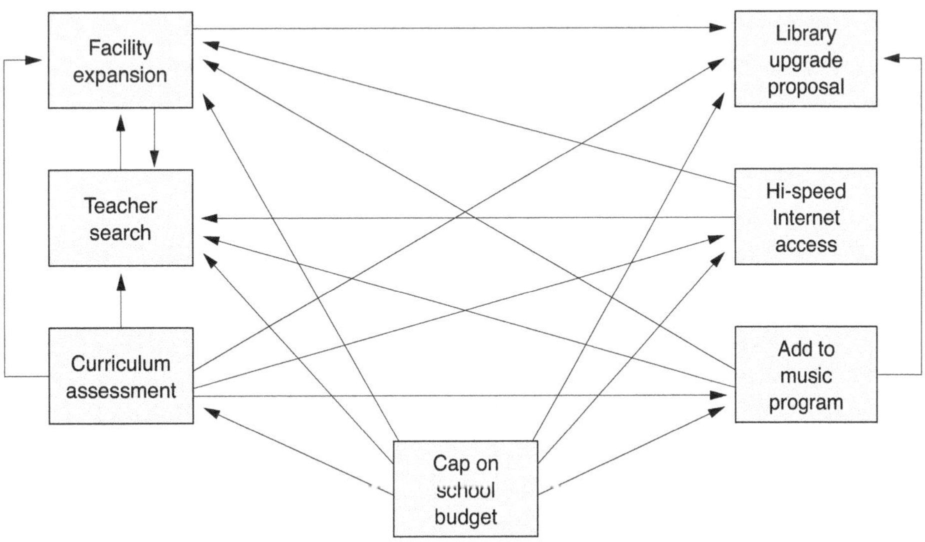

135. c; A *tree diagram* (also known as a *critical-to-quality tree*) is used to analyze and display subtopics and topics branching out from the main topic. This diagram (see figure below) is similar to an organization chart. Sometimes, it is flipped counterclockwise to read from left to right. The head of the organization is your main topic, subordinates and their staff and their organizations are

subtopics, and so forth. This diagram can be effectively used in a team setting to solicit inputs and map out the subtopics. An example of this application is root cause analysis of complex topics that can have multiple possibilities. [II.B.4] (Reference #3, Ch.7—B.4)

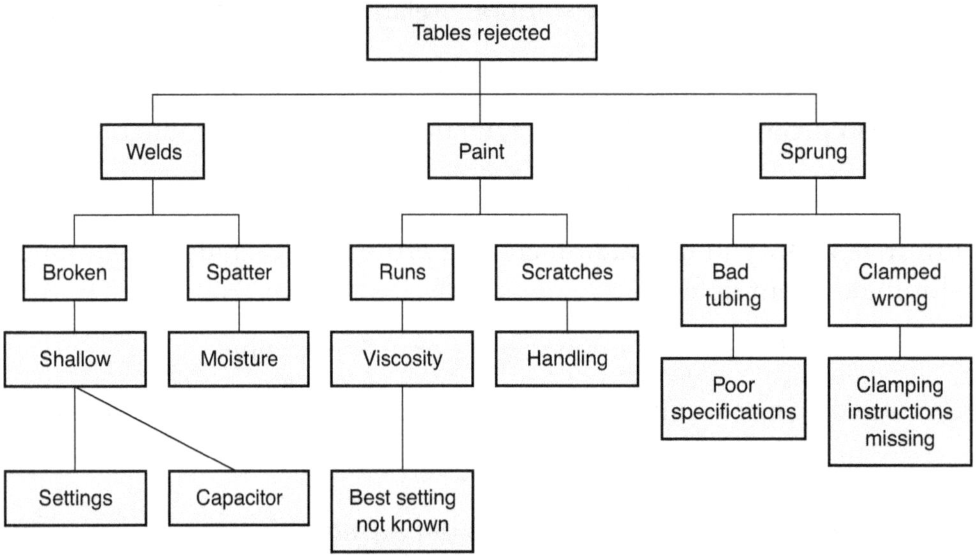

136. d; *Activity network diagram.* The *activity network diagram* (AND) is similar to the PERT chart because it graphically shows interdependencies between tasks. An AND highlights key tasks, the time to accomplish the tasks, flow paths (serial or parallel), and so on. This tool, like PERT and critical path analysis, can provide a top-level overview or detailed data depending on the project need. An example of an AND is shown below. [II.B.4] (Reference #2, Ch.7—D)

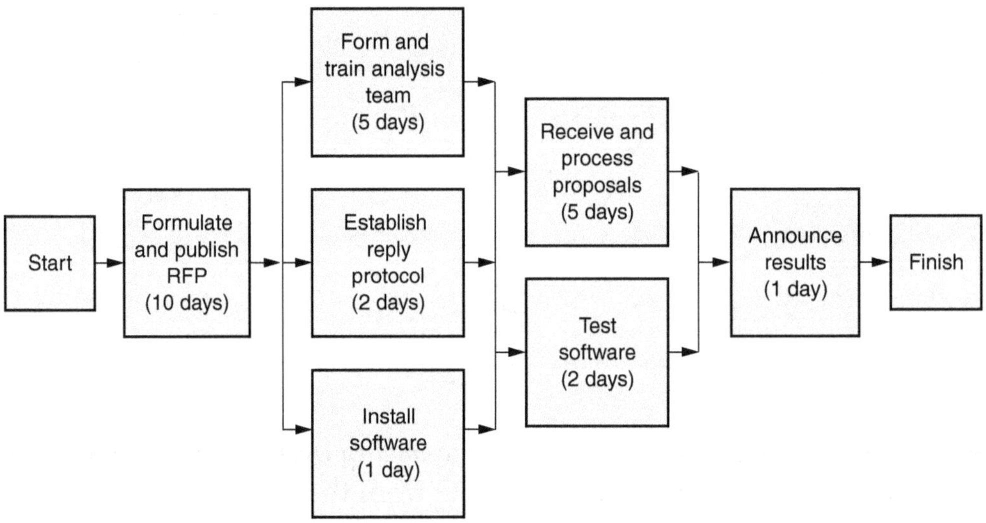

137. a; *Tollgate* or *phase reviews* are periodic progress reviews conducted for Six Sigma projects. The project manager and the project team may identify significant milestones at which the project should be reviewed. Milestones may be aligned with the natural stages of the Six Sigma structure (that is, define-measure-analyze-improve-control). Every organization is different. Some organizations set up entry and exit criteria for every stage to ensure that all the necessary due diligence is completed before moving to the next stage. There may also be significant milestones between stages.

As an example, recognizing the need for a Six Sigma project, building stakeholder consensus, and obtaining sponsor approval may likely be a milestone. Setting up a data collection system could be another milestone. The project manager and the team have to recognize the appropriate stages of the project in which to plan reviews. The purpose of this periodic review is to ensure that the project follows the necessary structure, deliverables are met, risks are anticipated and mitigated, project cost and time are closely monitored, and customers and other stakeholders are not negatively impacted. [II.B.5] (Reference #3, Ch.7—B.5)

138. b; *Tollgate* or *phase reviews* are periodic progress reviews conducted for Six Sigma projects. The project manager and the project team may identify significant milestones at which the project should be reviewed. Milestones may be aligned with the natural stages of the Six Sigma structure (that is, define-measure-analyze-improve-control). [II.B.5] (Reference #3, Ch.7—B.5)

139. c; As an example, recognizing the need for a Six Sigma project, building stakeholder consensus, and *obtaining sponsor approval may likely be a milestone.* Setting up a data collection system could be another milestone. The project manager and the team have to recognize the appropriate stages of the project in which to plan reviews. The purpose of this periodic review is to ensure that the project follows the necessary structure, deliverables are met, risks are anticipated and mitigated, project cost and time are closely monitored, and customer and other stakeholders are not negatively impacted. [II.B.5] (Reference #3, Ch.7—B.5)

Appendix A

Six Sigma Yellow Belt Sample Questions Cross-Referenced to Suggested Reference Material

CSSYB BoK category	Cognitive level	Title	Number of questions and question ID	Answer	Reference and page
I.A	Understand	Six Sigma foundations and principles	I.1	D	*CSSYB Handbook,* I.A
I.A	Understand	Six Sigma foundations and principles	I.2	A	*CSSGB Handbook,* Ch.1—A.1
I.A	Understand	Six Sigma foundations and principles	I.3	C	*Six Sigma for the Shop Floor,* The DMAIC Model
I.A	Understand	Six Sigma foundations and principles	I.4	D	*CSSGB Handbook,* Ch.1—A.1
I.A	Understand	Six Sigma foundations and principles	I.5	D	*CSSGB Handbook,* Ch.1—A.1
I.B	Understand	Lean foundations and principles	I.6	C	*CSSYB Handbook,* I.B
I.B	Understand	Lean foundations and principles	I.7	D	*CSSYB Handbook,* I.B
I.B	Understand	Lean foundations and principles	I.8	C	*Quality Improvement Handbook,* Ch.7—Lean Concepts and Tools
I.B	Understand	Lean foundations and principles	I.9	A	*CSSGB Handbook,* Ch.2—B
I.C	Understand	Six Sigma roles and responsibilities	I.10	B	*Quality Improvement Handbook,* Ch.5—Team Roles and Responsibilities
I.C	Understand	Six Sigma roles and responsibilities	I.11	A	*CSSYB Handbook,* I.C
I.C	Understand	Six Sigma roles and responsibilities	I.12	C	*CSSYB Handbook,* I.C
I.C	Understand	Six Sigma roles and responsibilities	I.13	D	*CSSYB Handbook,* I.C
I.D.1	Understand	Types of teams	I.14	C	*Quality Improvement Handbook,* Ch.4—Types of Teams
I.D.1	Understand	Types of teams	I.15	B	*Quality Improvement Handbook,* Ch.4—Types of Teams
I.D.1	Understand	Types of teams	I.16	C	*CSSGB Handbook,* Ch.9—F
I.D.2	Understand	Stages of development	I.17	D	*Team Handbook,* p. 6–7
I.D.2	Understand	Stages of development	I.18	D	*Team Handbook,* p. 3–52
I.D.2	Understand	Stages of development	I.19	A	*Team Handbook,* p. 6–5

CSSYB BoK category	Cognitive level	Title	Number of questions and question ID	Answer	Reference and page
I.D.2	Understand	Stages of development	I.20	C	*Team Handbook,* p. 6–7
I.D.3	Apply	Decision-making tools	I.21	C	*Quality Improvement Handbook,* Ch.6—Team Decision-Making
I.D.3	Apply	Decision-making tools	I.22	C	*Quality Toolbox,* Ch.5—The Tools—Brainstorming
I.D.3	Apply	Decision-making tools	I.23	B	*Quality Toolbox,* Ch.5—The Tools—Multivoting
I.D.3	Apply	Decision-making tools	I.24	B	*Quality Toolbox,* Ch.5—The Tools—Brainwriting
I.D.4	Apply	Communication methods	I.25	B	*Team Handbook,* p. 3–2
I.D.4	Apply	Communication methods	I.26	C	*CSSYB Handbook,* I.D.4
I.D.4	Apply	Communication methods	I.27	C	*CSSYB Handbook,* I.D.4
I.D.4	Apply	Communication methods	I.28	A	*Team Handbook,* p. 3–60
I.E.1	Apply	Quality tools	I.29	D	*Quality Toolbox,* Ch.5—The Tools—Flowchart
I.E.1	Apply	Quality tools	I.30	B	*Quality Improvement Handbook,* Ch.9—Improvement Tools
I.E.1	Apply	Quality tools	I.31	D	*Quality Improvement Handbook,* Ch.10—Root Cause Analysis
I.E.1	Apply	Quality tools	I.32	B	*Quality Toolbox,* Ch.5—The Tools—Run chart
I.E.1	Apply	Quality tools	I.33	C	*Quality Improvement Handbook,* Ch.9—Improvement Tools
I.E.1	Apply	Quality tools	I.34	D	*Quality Toolbox,* Ch.5—The Tools—Check Sheet
I.E.1	Apply	Quality tools	I.35	A	*Six Sigma for the Shop Floor,* Tools—Cause and Effect Diagrams
I.E.1	Apply	Quality tools	I.36	B	*CSSYB Handbook,* I.E.1
I.E.2	Apply	Six Sigma metrics	I.37	A	*CSSYB Handbook,* I.E.2
I.E.2	Apply	Six Sigma metrics	I.38	C	*CSSYB Handbook,* I.E.2
I.E.2	Apply	Six Sigma metrics	I.39	A	*CSSYB Handbook,* I.E.2
I.E.2	Apply	Six Sigma metrics	I.40	B	*CSSYB Handbook,* I.E.2
I.E.2	Apply	Six Sigma metrics	I.41	A	*CSSGB Handbook,* Ch.8—E.1
I.E.2	Apply	Six Sigma metrics	I.42	C	*Quality Toolbox,* Ch.5—The Tools—Cost of Poor Quality
II.A.1	Understand	Voice of the customer	II.1	B	*CSSYB Handbook,* II.A.1
II.A.1	Understand	Voice of the customer	II.2	C	*CSSYB Handbook,* II.A.1
II.A.1	Understand	Voice of the customer	II.3	C	*CSSGB Handbook,* Ch.5—B.1
II.A.2	Understand	Project selection	II.4	A	*CSSGB Handbook,* Ch.1—A.2

CSSYB BoK category	Cognitive level	Title	Number of questions and question ID	Answer	Reference and page
II.A.2	Understand	Project selection	II.5	B	*CSSYB Handbook*, II.A.2
II.A.2	Understand	Project selection	II.6	A	*CSSGB Handbook*, Ch.4—A.1
II.A.3	Understand	Stakeholder analysis	II.7	C	*CSSGB Handbook*, Ch.4—A.5
II.A.3	Understand	Stakeholder analysis	II.8	B	*CSSGB Handbook*, Ch.4—A.5; *Quality Improvement Handbook*, Ch.8—Cost of Poor Quality.
II.A.3	Understand	Stakeholder analysis	II.9	A	*Quality Improvement Handbook*, Ch.15—Customer Identification
II.A.4	Apply	Process inputs and outputs	II.10	D	*Quality Improvement Handbook*, Ch.1—Terms, Concepts, and Principles
II.A.4	Apply	Process inputs and outputs	II.11	C	*Quality Toolbox*, Ch.5—The Tools—SIPOC
II.A.4	Apply	Process inputs and outputs	II.12	D	*CSSYB Handbook*, II.A.4
II.A.5	Understand	Supply Chain Management	II.13	B	*Quality Improvement Handbook*, Appendix C
II.A.5	Understand	Supply Chain Management	II.14	C	*Quality Improvement Handbook*, Ch. 13—Supplier Relationships
II.B.1	Understand	Project charter	II.15	A	*CSSGB Handbook*, Ch. 6—C.1
II.B.1	Understand	Project charter	II.16	C	*CSSYB Handbook*, II.B.1
II.B.2	Understand	Communication plan	II.17	A	*CSSYB Handbook*, II.B.2
II.B.2	Understand	Communication plan	II.18	B	*CSSYB Handbook*, II.B.2
II.B.3	Understand	Project planning	II.19	B	*CSSYB Handbook*, II.B.3
II.B.3	Understand	Project planning	II.20	C	*CSSGB Handbook*, Ch.7—D
II.B.3	Understand	Project planning	II.21	A	*Six Sigma for the Shop Floor*, Tools—Project Planning
II.B.4	Understand	Project management tools	II.22	D	*Quality Toolbox*, Ch.5—The Tools—Affinity Diagraming
II.B.4	Understand	Project management tools	II.23	D	*CSSGB Handbook*, Ch.7—D; *CSSYB Handbook*, II.B.4
II.B.4	Understand	Project management tools	II.24	A	*CSSGB Handbook*, Ch.7—D
II.B.5	Understand	Phase reviews	II.25	C	*CSSYB Handbook*, II.B.5
II.B.5	Understand	Phase reviews	II.26	C	*CSSYB Handbook*, II.B.5
III.A	Apply	Basic statistics	III.1	B	*CSSGB Handbook*, Ch.13—D.3; *Quality Improvement Handbook*, Ch.9—Quality Improvement Tools
III.A	Apply	Basic statistics	III.2	A	*CSSYB Handbook*, III.A
III.A	Apply	Basic statistics	III.3	C	*CSSYB Handbook*, III.A
III.A	Apply	Basic statistics	III.4	C	*CSSYB Handbook*, III.A

CSSYB BoK category	Cognitive level	Title	Number of questions and question ID	Answer	Reference and page
III.A	Apply	Basic statistics	III.5	A	*CSSYB Handbook*, III.A; *Six Sigma for the Shop Floor*, Tools—Histogram
III.A	Apply	Basic statistics	III.6	C	*CSSYB Handbook*, III.A
III.B.1	Understand	Data collection plans	III.7	D	*CSSGB Handbook*, Ch.13—D.1
III.B.1	Understand	Data collection plans	III.8	A	*CSSYB Handbook*, III.B.1
III.B.1	Understand	Data collection plans	III.9	C	*CSSYB Handbook*, III.B.1
III.B.1	Understand	Data collection plans	III.10	B	*CSSYB Handbook*, III.B.1
III.B.1	Understand	Data collection plans	III.11	D	*CSSYB Handbook*, III.B.1
III.B.2	Understand	Qualitative and quantitative data	III.12	C	*Quality Improvement Handbook*, Ch.16—Voice of the Customer; *CSSGB Handbook*, Ch.13—D.1
III.B.2	Understand	Qualitative and quantitative data	III.13	A	*CSSGB Handbook*, Ch.13—D.1; *CSSYB Handbook*, III.B.2
III.B.2	Understand	Qualitative and quantitative data	III.14	D	*CSSYB Handbook*, III.B.2
III.B.2	Understand	Qualitative and quantitative data	III.15	C	*CSSGB Handbook*, Ch.21—A.3
III.B.2	Understand	Qualitative and quantitative data	III.16	A	*CSSYB Handbook*, III.B.2
III.B.2	Understand	Qualitative and quantitative data	III.17	B	*CSSGB Handbook*, Ch.13—D.1
III.B.3	Apply	Data collection techniques	III.18	C	*CSSYB Handbook*, III.B.3
III.B.3	Apply	Data collection techniques	III.19	B	*CSSYB Handbook*, III.B.3
III.B.3	Apply	Data collection techniques	III.20	C	*CSSYB Handbook*, III.B.3
III.B.3	Apply	Data collection techniques	III.21	D	*CSSYB Handbook*, III.B.3
III.B.3	Apply	Data collection techniques	III.22	A	*Quality Toolbox*, Ch.5—The Tools—E-Survey
III.C.1	Understand	MSA terms	III.23	C	*CSSYB Handbook*, III.C.1
III.C.1	Understand	MSA terms	III.24	D	*CSSYB Handbook*, III.C.1
III.C.1	Understand	MSA terms	III.25	C	*CSSYB Handbook*, III.C.1
III.C.1	Understand	MSA terms	III.26	A	*CSSYB Handbook*, III.C.1
III.C.1	Understand	MSA terms	III.27	C	*CSSYB Handbook*, III.C.1
III.C.2	Understand	Gauge repeatability & reproducibility	III.28	A	*Quality Toolbox*, p. 448
III.C.2	Understand	Gauge repeatability & reproducibility	III.29	D	*CSSGB Handbook*, Ch.14—E; *CSSYB Handbook*, III.C.2
III.C.2	Understand	Gauge repeatability & reproducibility	III.30	A	*CSSGB Handbook*, Ch.14—E; *CSSYB Handbook*, III.C.2
IV.A.1	Understand	Process analysis tools	IV.1	C	*CSSGB Handbook*, Ch.2—B.1

CSSYB BoK category	Cognitive level	Title	Number of questions and question ID	Answer	Reference and page
IV.A.1	Understand	Process analysis tools	IV.2	A	*CSSYB Handbook,* I.A.1
IV.A.1	Understand	Process analysis tools	IV.3	A	*Quality Improvement Handbook,* Ch.7—Process Improvement
IV.A.1	Understand	Process analysis tools	IV.4	B	*Quality Improvement Handbook,* Appendix C—Quality Glossary
IV.A.2	Apply	Failure mode and effect analysis	IV.5	B	*Six Sigma for the Shop Floor,* Tools—Failure Mode and Effects Analysis
IV.A.2	Apply	Failure mode and effect analysis	IV.6	D	*CSSGB Handbook,* Ch.3—C.2
IV.A.2	Apply	Failure mode and effect analysis	IV.7	C	*CSSGB Handbook,* Ch.3—C.2
IV.A.2	Apply	Failure mode and effect analysis	IV.8	A	*CSSYB Handbook,* IV.A.2t
IV.B	Understand	Root cause analysis	IV.9	C	*Quality Improvement Handbook,* Ch.9—Improvement Tools; *CSSYB Handbook,* IV.B
IV.B	Understand	Root cause analysis	IV.10	D	*CSSYB Handbook,* IV.B
IV.B	Understand	Root cause analysis	IV.11	A	*CSSGB Handbook,* Ch.9—B
IV.B	Understand	Root cause analysis	IV.12	B	*Quality Improvement Handbook,* Ch.10—Root Cause Analysis
IV.B	Understand	Root cause analysis	IV.13	C	*Six Sigma for the Shop Floor,* Tools—Problem Solving
IV.B	Understand	Root Cause Analysis	IV.14	B	*Quality Improvement Handbook,* Appendix C—Quality Glossary
IV.C	Apply	Corrective action	IV.15	C	*CSSGB Handbook,* Ch.19—B
IV.C	Apply	Corrective action	IV.16	A	*Six Sigma for the Shop Floor,* Tools—Problem Solving
IV.D	Apply	Preventive action	IV.17	D	*Quality Improvement Handbook,* Ch.7—Process Improvement
IV.D	Apply	Preventive action	IV.18	C	*Quality Improvement Handbook,* Ch.9—Improvement Tools
IV.E.1	Understand	Basic distribution types	IV.19	A	*CSSGB Handbook,* Ch.12—C
IV.E.1	Understand	Basic distribution types	IV.20	C	*Quality Toolbox,* Ch.5—The Tools—Histograms and other Frequency Distributions
IV.E.1	Understand	Basic distribution types	IV.21	C	*Quality Toolbox,* Ch.5—The Tools—Histograms and other Frequency Distributions
IV.E.1	Understand	Basic distribution types	IV.22	B	*CSSGB Handbook,* Ch.2—C
IV.E.1	Understand	Basic distribution types	IV.23	B	*CSSYB Handbook,* IV.E.1
IV.E.2	Understand	Common and special cause variation	IV.24	D	*Quality Improvement Handbook,* Ch.9—Improvement Tools

CSSYB BoK category	Cognitive level	Title	Number of questions and question ID	Answer	Reference and page
IV.E.2	Understand	Common and special cause variation	IV.25	A	*Quality Improvement Handbook,* Ch.9—Improvement Tools
IV.E.2	Understand	Common and special cause variation	IV.26	C	*CSSYB Handbook,* IV.E.2
IV.F.1	Understand	Correlation	IV.27	B	*CSSY Handbook,* IV.F.1; *Quality Toolbox,* Ch.5—The Tools—Correlation Analysis
IV.F.1	Understand	Correlation	IV.28	A	*CSSGB Handbook,* Ch.16—A.2
IV.F.1	Understand	Correlation	IV.29	A	*CSSYB Handbook,* IV.F.1
IV.F.2	Understand	Regression	IV.30	B	*Quality Toolbox,* Ch.5—The Tools—Regression Analysis
IV.F.2	Understand	Regression	IV.31	D	*CSSGB Handbook,* Ch.16—A.2
IV.F.2	Understand	Regression	IV.32	B	*Quality Toolbox,* Ch.5—The Tools—Regression Analysis
IV.F.2	Understand	Regression	IV.33	A	*CSSYB Handbook,* IV.F.2
IV.G	Understand	Hypothesis testing	IV.34	A	*CSSGB Handbook,* Ch.17—B.1
IV.G	Understand	Hypothesis testing	IV.35	C	*CSSYB Handbook,* IV.G
IV.G	Understand	Hypothesis testing	IV.36	A	*CSSYB Handbook,* IV.G
IV.G	Understand	Hypothesis testing	IV.37	A	*CSSYB Handbook,* IV.G
V.A.1	Understand	Improvement techniques	V.1	D	*Quality Improvement Handbook,* Ch.7—Process Improvement
V.A.1	Understand	Improvement techniques	V.2	B	*Quality Improvement Handbook,* Ch.7—Process Improvement
V.A.1	Understand	Improvement techniques	V.3	A	*CSSYB Handbook,* V.A.1
V.A.1	Understand	Improvement techniques	V.4	B	*Quality Improvement Handbook,* Ch.7—Process Improvement
V.A.2	Understand	Plan-do-check-act (PDCA) cycle	V.5	C	*Quality Improvement Handbook,* Ch.8—Improvement Techniques
V.A.2	Understand	Plan-do-check-act (PDCA) cycle	V.6	A	*CSSYB Handbook,* V.A.2
V.A.2	Understand	Plan-do-check-act (PDCA) cycle	V.7	B	*Quality Improvement Handbook,* Ch.8—Improvement Techniques
V.A.2	Understand	Plan-do-check-act (PDCA) cycle	V.8	A	*CSSYB Handbook,* V.A.2
V.A.3	Understand	Cost/benefit analysis	V.9	A	*CSSYB Handbook,* V.A.3
V.A.3	Understand	Cost/benefit analysis	V.10	C	*CSSYB Handbook,* V.A.3
V.A.3	Understand	Cost/benefit analysis	V.11	C	*CSSYB Handbook,* V.A.3
V.B.1	Understand	Control plan	V.12	D	*Six Sigma for the Shop Floor,* Tools—Control Plan
V.B.1	Understand	Control plan	V.13	B	*CSSGB Handbook,* Ch.22—B
V.B.1	Understand	Control plan	V.14	A	*CSSYB Handbook,* V.B.1
V.B.1	Understand	Control plan	V.15	A	*CSSYB Handbook,* V.B.1

CSSYB BoK category	Cognitive level	Title	Number of questions and question ID	Answer	Reference and page
V.B.1	Understand	Control plan	V.16	C	*CSSYB Handbook,* V.B.1
V.B.2	Understand	Control charts	V.17	A	*CSSYB Handbook,* V.B.2
V.B.2	Understand	Control charts	V.18	C	*Quality Toolbox,* Ch.5—The Tools—Control Chart
V.B.2	Understand	Control charts	V.19	B	*Quality Toolbox,* Ch.5—The Tools—Control Chart
V.B.2	Understand	Control charts	V.20	A	*CSSGB Handbook,* Ch.21—A.3
V.B.2	Understand	Control charts	V.21	B	*CSSGB Handbook,* Ch.21—A.1
V.B.3	Understand	Document control	V.22	D	*CSSYB Handbook,* V.B.3
V.B.3	Understand	Document control	V.23	A	*CSSYB Handbook,* V.B.3
V.B.3	Understand	Document control	V.24	C	*CSSYB Handbook,* V.B.3
V.B.4	Understand	Work instructions and SOPs	V.25	C	*CSSGB Handbook,* Ch.10—A
V.B.4	Understand	Work instructions and SOPs	V.26	B	*CSSGB Handbook,* Ch.10—A

Appendix B

Six Sigma Yellow Belt Additional Questions Cross-Referenced to Suggested Reference Material

CSSYB BoK category	Cognitive level	Title	Number of questions and question ID	Answer	Reference and page
II.B.1	Understand	Project charter	A.1	B	*Team Handbook,* p. 2–15
I.D.1	Understand	Types of teams	A.2	B	*Quality Improvement Handbook,* Ch.6—Initiating Teams
I.D.1	Understand	Types of teams	A.3	D	*Quality Improvement Handbook,* Ch.4—Types of Teams
I.D.1	Understand	Types of teams	A.4	C	*Quality Improvement Handbook,* Ch.6—Initiating Teams
I.D.1	Understand	Types of teams	A.5	C	*Team Handbook,* p. 2–15
I.D.2	Understand	Types of teams	A.6	A	*CSSYB Handbook,* I.D.2
I.D.2	Understand	Types of teams	A.7	A	*Team Handbook,* p. 6–6
I.D.2	Understand	Types of teams	A.8	D	*Team Handbook,* p. 6–4
I.D.2	Understand	Types of teams	A.9	C	*Team Handbook,* p. 6–9
I.E.1	Apply	Quality tools	A.10	B	*Quality Toolbox,* Ch.5—Top-Down Flowchart
I.E.1	Apply	Quality tools	A.11	D	*Quality Toolbox,* Ch.5—Deployment Flowchart
I.E.1	Apply	Quality tools	A.12	A	*Quality Toolbox, Ch.5—Histogram*
I.E.1	Apply	Quality tools	A.13	D	*Quality Toolbox,* Ch.5—Pareto Chart
I.E.1	Apply	Quality tools	A.14	A	*CSSGB Handbook,* Ch.7—D
II.A.3	Understand	Stakeholder analysis	A.15	D	*CSSGB Handbook,* Ch.4—A.5
II.A.5	Understand	Supply Chain Management	A.16	A	*Quality Improvement Associate Handbook,* Ch.13—The Process of Supply Chains
II.A.5	Understand	Supply Chain Management	A.17	D	*Quality Improvement Associate Handbook,* Ch.13—Supplier Relationship
II.B.4	Understand	Project management tools	A.18	D	*CSSYB Handbook,* II.B.4
IV.E.2	Understand	Common and special cause variation	A.19	B	*CSSGB Handbook,* Ch.1—A.1; *Quality Improvement Handbook,* Glossary
V.B.2	Understand	Control charts	A.20	C	*Quality Improvement Handbook,* Ch.9—Control Chart

CSSYB BoK category	Cognitive level	Title	Number of questions and question ID	Answer	Reference and page
III.A	Apply	Basic statistics	A.21	A	*CSSYB Handbook,* III.A
III.A	Apply	Basic statistics	A.22	A	*CSSYB Handbook,* III.A
III.A	Apply	Basic statistics	A.23	D	*CSSYB Handbook,* III.A
III.B.1	Understand	Data collection plans	A.24	C	*CSSYB Handbook,* III.B
III.B.1	Understand	Data collection plans	A.25	B	*CSSYB Handbook,* III.B.1
III.B.2	Understand	Qualitative and quantitative data	A.26	C	*CSSGB Handbook,* Ch.3—D.1; *CSSYB Handbook,* III.B.2
III.B.2	Understand	Qualitative and quantitative data	A.27	A	*CSSGB Handbook,* Ch.3—D.1
III.B.3	Apply	Data collection techniques	A.28	B	*Quality Toolbox,* Ch.5—Telephone Interview
III.B.3	Apply	Data collection techniques	A.29	C	*Quality Toolbox,* Ch.5—Face-to-Face Interview
III.B.3	Apply	Data collection techniques	A.30	D	*Quality Toolbox,* Ch.5—Question-naire
III.C.1	Understand	MSA terms	A.31	B	*CSSYB Handbook,* III.C.1
III.C.1	Understand	MSA terms	A.32	B	*CSSYB Handbook,* III.C.1
III.C.1	Understand	MSA terms	A.33	B	*CSSYB Handbook,* III.C.1, Glossary
III.C.2	Understand	Gauge repeatability & reproducibility	A.34	A	*Six Sigma for the Shop Floor,* Tools—MSA; *Quality Toolbox,* Ch.5—Repeatability
III.C.2	Understand	Gauge repeatability & reproducibility	A.35	A	*Quality Toolbox,* Ch.5—Repeatability & Reproducibility Study
III.C.2	Understand	Gauge repeatability & reproducibility	A.36	B	*Quality Toolbox,* Ch.5—Repeatability & Reproducibility Study
IV.A.1	Understand	Lean tools	A.37	D	*Quality Improvement Handbook,* Ch.7—How to Start the Lean Journey
IV.A.2	Apply	Failure mode and effect analysis	A.38	C	*CSSYB Handbook,* IV.A.2
IV.A.2	Apply	Failure mode and effect analysis	A.39	B	*CSSYB Handbook,* IV.A.2
IV.A.2	Apply	Failure mode and effect analysis	A.40	B	*CSSYB Handbook,* IV.A.2
IV.B	Understand	Root cause analysis	A.41	B	*CSSGB Handbook,* Ch.19—B Cause & Effect (x—y) Relational Matrix
IV.B	Understand	Root cause analysis	A.42	A	*CSSYB Handbook,* IV.B
IV.E.1	Understand	Basic distribution types	A.43	D	*CSSYB Handbook,* IV.C.1
IV.E.1	Understand	Basic distribution types	A.44	B	*CSSYB Handbook,* IV.C.1
IV.E.1	Understand	Basic distribution types	A.45	A	*CSSYB Handbook,* IV.C.1

CSSYB BoK category	Cognitive level	Title	Number of questions and question ID	Answer	Reference and page
IV.E.2	Understand	Common and special cause variation	A.46	D	*CSSYB Handbook,* IV.C.1, Glossary
IV.E.2	Understand	Common and special cause variation	A.47	C	*CSSYB Handbook,* IV.C.2
IV.E.2	Understand	Common and special cause variation	A48	B	*CSSYB Handbook,* IV.E.2
IV.F.1	Understand	Correlation	A49	A	*CSSYB Handbook,* IV.F.1
IV.F.1	Understand	Correlation	A.50	B	*CSSYB Handbook,* IV.F.1
IV.F.1	Understand	Correlation	A.51	A	*CSSYB Handbook,* IV.F.1
IV.F.2	Understand	Regression	A.52	C	*CSSYB Handbook,* IV.F.2
IV.F.2	Understand	Regression	A.53	B	*CSSYB Handbook,* IV.F.2
IV.F.2	Understand	Regression	A54	D	*Quality Toolbox,* Ch.5—Regression Analysis
IV.G	Understand	Hypothesis testing	A.55	B	*CSSYB Handbook,* IV.G
IV.G	Understand	Hypothesis testing	A.56	A	*CSSYB Handbook,* IV.G; *CSSGB Handbook,* Ch.17—B.1
V.A.1	Understand	Kaizen and kaizen blitz	A.57	B	*CSSGB Handbook,* Ch.2—B.1
V.A.1	Understand	Kaizen and kaizen blitz	A.58	A	*CSSGB Handbook,* Ch.2—B.1
V.A.1	Understand	Kaizen and kaizen blitz	A.59	A	*Quality Improvement Handbook,* Ch.7—Kaizen
V.A.2	Understand	Plan-do-check-act (PDCA)	A.60	C	*Six Sigma for the Shop Floor,* Tools—PDSA
V.A.2	Understand	Plan-do-check-act (PDCA)	A.61	C	*CSSYB Handbook,* V.A.2
V.A.3	Understand	Cost-benefit analysis	A.62	B	*CSSYB Handbook,* V.A.3
V.A.3	Understand	Cost-benefit analysis	A.63	A	*CSSYB Handbook,* V.A.3
V.A.3	Understand	Cost-benefit analysis	A.64	C	*CSSYB Handbook,* V.A.3
V.B.1	Understand	Control plan	A.65	D	*CSSYB Handbook,* V.B.1
V.B.2	Understand	Control charts	A.66	B	*CSSYB Handbook,* V.B.2
V.B.2	Understand	Control charts	A.67	D	*CSSYB Handbook,* V.B.2
V.B.2	Understand	Control charts	A.68	D	*Quality Toolbox,* Ch.5—Run Chart
V.B.3	Understand	Document control	A.69	B	*CSSYB Handbook,* V.B.3
V.B.4	Understand	Work instructions and SOPs	A.70	B	*Six Sigma for the Shop Floor,* Tools—SOPs
I.A	Understand	Six Sigma foundations and principles	A.71	C	*Six Sigma for the Shop Floor,* What is Six Sigma?
I.A	Understand	Six Sigma foundations and principles	A.72	B	*Quality Improvement Handbook,* Ch.7—The DMAIC Methodology
I.A	Understand	Six Sigma foundations and principles	A.73	D	*CSSGB Handbook,* Ch.1—A.1

CSSYB BoK category	Cognitive level	Title	Number of questions and question ID	Answer	Reference and page
I.A	Understand	Six Sigma foundations and principles	A.74	A	*Quality Improvement Handbook,* Ch.7—Six Sigma Concepts and Tools
I.B	Understand	Lean foundations and principles	A.75	B	*CSSYB Handbook,* I.B
I.B	Understand	Lean foundations and principles	A.76	A	*CSSYB Handbook,* I.B
I.B	Understand	Lean foundations and principles	A.77	C	*Quality Improvement Handbook,* Ch.7—Lean Concepts and Tools
I.B	Understand	Lean Foundations and Principles	A.78	A	*CSSGB Handbook,* Ch.2–B
I.C	Understand	Six Sigma roles and responsibilities	A.79	D	*CSSYB Handbook,* I.C t
I.C	Understand	Six Sigma roles and responsibilities	A.80	B	*Quality Improvement Handbook,* Ch.7—The DMAIC Methodology t
I.C	Understand	Six Sigma roles and responsibilities	A.81	D	*CSSYB Handbook,* I.C t
I.C	Understand	Six Sigma roles and responsibilities	A.82	A	*Quality Improvement Handbook,* Ch.7—The DMAIC Methodology t
I.D.1	Understand	Types of teams	A.83	B	*CSSGB Handbook,* Ch.9—F
I.D.1	Understand	Types of teams	A.84	C	*CSSGB Handbook,* Ch.9—F
I.D.1	Understand	Types of teams	A.85	D	*Six Sigma for the Shop Floor,* Tools—Self-Directed Teams
I.D.1	Understand	Types of teams	A.86	B	*CSSYB Handbook,* I.D.1
I.D.2	Understand	Stages of development	A.87	C	*Team Handbook,* p. 6–2
I.D.2	Understand	Stages of development	A.88	D	*Team Handbook,* p. 6–8
I.D.2	Understand	Stages of development	A.89	B	*Quality Improvement Handbook,* Ch.6—Team Stages
I.D.2	Understand	Stages of development	A.90	A	*Team Handbook,* p. 6–4
I.D.3	Apply	Decision-making tools	A.91	B	*Quality Toolbox,* Ch.5—Multivoting
I.D.3	Apply	Decision-making tools	A.92	C	*Quality Improvement Handbook,* Ch.6—Team Decision-Making
I.D.3	Apply	Decision-making tools	A.93	D	*CSSYB Handbook,* I.D.3
I.D.3	Apply	Decision-making tools	A.94	B	*Quality Toolbox,* Ch.5—Brainwriting
I.D.4	Apply	Communication methods	A.95	C	*CSSYB Handbook,* I.D.4
I.D.4	Apply	Communication methods	A.96	D	*Six Sigma for the Shop Floor,* Tools—Team Meetings
I.D.4	Apply	Communication methods	A.97	A	*Team Handbook,* p. 3–60
I.D.4	Apply	Communication methods	A.98	B	*Quality Improvement Handbook,* Ch.6—Other Barriers
I.E.1	Apply	Quality tools	A.99	C	*Handbook,* Ch.9—Scatter Diagram

CSSYB BoK category	Cognitive level	Title	Number of questions and question ID	Answer	Reference and page
I.E.1	Apply	Quality tools	A.100	D	*Quality Toolbox,* Ch.5—Check Sheet
I.E.1	Apply	Quality tools	A.101	A	*Six Sigma for the Shop Floor,* Tools—Cause and Effect Diagram
I.E.1	Apply	Quality tools	A.102	B	*CSSYB Handbook,* I.E.1
I.E.2	Apply	Six Sigma metrics	A.103	C	*Quality Toolbox,* Ch.5—Cost of Poor Quality Analysis
I.E.2	Apply	Six Sigma metrics	A.104	D	*Quality Improvement Handbook,* Ch.8—Cost of Poor Quality or Cost of Quality
I.E.2	Apply	Six Sigma metrics	A.105	A	*CSSYB Handbook,* I.E.2
I.E.2	Apply	Six Sigma metrics	A.106	B	*Quality Improvement Handbook,* Ch.7—Cycle Time Reduction
II.A.1	Understand	Voice of the customer	A.107	C	*CSSGB Handbook,* Ch.5—B.1
II.A.1	Understand	Voice of the customer	A.108	D	*Quality Improvement Handbook,* Ch.7—The DMAIC Methodology
II.A.1	Understand	Voice of the customer	A.109	B	*Quality Toolbox,* Tools—Stakeholder Analysis
II.A.1	Understand	Voice of the customer	A.110	C	*CSSYB Handbook,* II.A.2
II.A.2	Understand	Project selection	A.111	A	*CSSGB Handbook,* Ch.4—A.1
II.A.2	Understand	Project selection	A.112	D	*CSSGB Handbook,* Ch.4—A.1
II.A.2	Understand	Project selection	A.113	B	*CSSYB Handbook,* II.A.2
II.A.2	Understand	Project selection	A.114	D	*Quality Improvement Handbook,* Ch.7—The DMAIC Methodology
II.A.3	Understand	Stakeholder analysis	A.115	C	*CSSYB Handbook,* II.A.3
II.A.3	Understand	Stakeholder analysis	A.116	B	*CSSYB Handbook,* II.A.3
II.A.3	Understand	Stakeholder analysis	A.117	A	*Quality Improvement Handbook,* Glossary
II.A.3	Understand	Stakeholder analysis	A.118	C	*CSSGB Handbook,* Ch.4—A.5
II.A.4	Apply	Process inputs and outputs	A.119	D	*CSSYB Handbook,* II.A.4
II.A.4	Apply	Process inputs and outputs	A.120	A	*Quality Toolbox,* Tools—SIPOC Diagram
II.A.4	Apply	Process inputs and outputs	A.121	C	*Quality Improvement Handbook,* Ch.1—Taking a Systems View of Improvement
II.A.4	Apply	Process inputs and outputs	A.122	D	*CSSYB Handbook,* II.A.4
II.B.1	Understand	Project charter	A.123	A	*CSSGB Handbook,* Ch.6—A.4
II.B.1	Understand	Project charter	A.124	B	*Quality Improvement Handbook,* Glossary
II.B.1	Understand	Project charter	A.125	C	*CSSYB Handbook,* II.B.1
II.B.1	Understand	Project charter	A.126	D	*Team Handbook,* p. 3–48
II.B.2	Understand	Communication plan	A.127	A	*Team Handbook,* p. 1–13

CSSYB BoK category	Cognitive level	Title	Number of questions and question ID	Answer	Reference and page
II.B.2	Understand	Communication plan	A.128	B	*CSSYB Handbook,* Ch.7—B.2
II.B.3	Understand	Project planning	A.129	A	*Six Sigma for the Shop Floor,* Tools—Project Planning
II.B.3	Understand	Project planning	A.130	B	*Quality Improvement Handbook,* Glossary
II.B.3	Understand	Project planning	A.131	C	*CSSYB Handbook,* II.B.3
II.B.3	Understand	Project planning	A.132	D	*Quality Toolbox,* Tools—Gantt Chart
II.B.4	Understand	Project management tools	A.133	A	*CSSGB Handbook,* Ch.7—D
II.B.4	Understand	Project management tools	A.134	B	*Quality Improvement Handbook,* Ch.9—Relations Diagram
II.B.4	Understand	Project management tools	A.135	C	*CSSYB Handbook,* Ch.7—B.4
II.B.4	Understand	Project management tools	A.136	D	*CSSGB Handbook,* Ch.7—D
II.B.5	Understand	Phase reviews	A.137	A	*CSSYB Handbook,* Ch.7—B.5
II.B.5	Understand	Phase reviews	A.138	B	*CSSYB Handbook,* Ch.7—B.5
II.B.5	Understand	Phase reviews	A.139	C	*CSSYB Handbook,* Ch.7—B.5

Appendix C

Sigma Level Conversion Table

With no sigma shift (centered)				With 1.5 sigma shift			
Sigma level	Percent in specification	Percent defective	ppm	Sigma level	Percent in specification	Percent defective	ppm
0.10	7.9656	92.0344	920344	0.10	2.5957	97.40426	974043
0.20	15.8519	84.1481	841481	0.20	5.2235	94.77650	947765
0.30	23.5823	76.4177	764177	0.30	7.9139	92.08606	920861
0.40	31.0843	68.9157	689157	0.40	10.6950	89.30505	893050
0.50	38.2925	61.7075	617075	0.50	13.5905	86.40949	864095
0.60	45.1494	54.8506	548506	0.60	16.6196	83.38043	833804
0.70	51.6073	48.3927	483927	0.70	19.7952	80.20480	802048
0.80	57.6289	42.3711	423711	0.80	23.1240	76.87605	768760
0.90	63.1880	36.8120	368120	0.90	26.6056	73.39444	733944
1.00	68.2689	31.7311	317311	1.00	30.2328	69.76721	697672
1.10	72.8668	27.1332	271332	1.10	33.9917	66.00829	660083
1.20	76.9861	23.0139	230139	1.20	37.8622	62.13784	621378
1.30	80.6399	19.3601	193601	1.30	41.8185	58.18148	581815
1.40	83.8487	16.1513	161513	1.40	45.8306	54.16937	541694
1.50	86.6386	13.3614	133614	1.50	49.8650	50.13499	501350
1.60	89.0401	10.9599	109599	1.60	53.8860	46.11398	461140
1.70	91.0869	8.9131	89131	1.70	57.8573	42.14274	421427
1.80	92.8139	7.1861	71861	1.80	61.7428	38.25720	382572
1.90	94.2567	5.7433	57433	1.90	65.5085	34.49152	344915
2.00	95.4500	4.5500	45500	2.00	69.1230	30.87702	308770
2.10	96.4271	3.5729	35729	2.10	72.5588	27.44122	274412
2.20	97.2193	2.7807	27807	2.20	75.7929	24.20715	242071
2.30	97.8552	2.1448	21448	2.30	78.8072	21.19277	211928
2.40	98.3605	1.6395	16395	2.40	81.5892	18.41082	184108
2.50	98.7581	1.2419	12419	2.50	84.1313	15.86869	158687
2.60	99.0678	0.9322	9322	2.60	86.4313	13.56867	135687
2.70	99.3066	0.6934	6934	2.70	88.4917	11.50830	115083
2.80	99.4890	0.5110	5110	2.80	90.3191	9.68090	96809

With no sigma shift (centered)				With 1.5 sigma shift			
Sigma level	Percent in specification	Percent defective	ppm	Sigma level	Percent in specification	Percent defective	ppm
2.90	99.6268	0.3732	3732	2.90	91.9238	8.07621	80762
2.90	99.6268	0.3732	3732	2.90	91.9238	8.07621	80762
3.00	99.7300	0.2700	2700	3.00	93.3189	6.68106	66811
3.10	99.8065	0.1935	1935	3.10	94.5199	5.48014	54801
3.20	99.8626	0.1374	1374	3.20	95.5433	4.45668	44567
3.30	99.9033	0.0967	967	3.30	96.4069	3.59311	35931
3.40	99.9326	0.0674	674	3.40	97.1283	2.87170	28717
3.50	99.9535	0.0465	465	3.50	97.7250	2.27504	22750
3.60	99.9682	0.0318	318	3.60	98.2135	1.78646	17865
3.70	99.9784	0.0216	216	3.70	98.6096	1.39035	13904
3.80	99.9855	0.0145	145	3.80	98.9276	1.07242	10724
3.90	99.9904	0.0096	96.2	3.90	99.1802	0.81976	8198
4.00	99.9937	0.0063	63.3	4.00	99.3790	0.62097	6210
4.10	99.9959	0.0041	41.3	4.10	99.5339	0.46612	4661
4.20	99.9973	0.0027	26.7	4.20	99.6533	0.34670	3467
4.30	99.9983	0.0017	17.1	4.30	99.7445	0.25551	2555
4.40	99.9989	0.0011	10.8	4.40	99.8134	0.18658	1866
4.50	99.9993	0.0007	6.8	4.50	99.8650	0.13499	1350
4.60	99.9996	0.0004	4.2	4.60	99.9032	0.09676	968
4.70	99.9997	0.0003	2.6	4.70	99.9313	0.06871	687
4.80	99.9998	0.0002	1.6	4.80	99.9517	0.04834	483
4.90	99.99990	0.00010	1.0	4.90	99.9663	0.03369	337
5.00	99.99994	0.00006	0.6	5.00	99.9767	0.02326	233
5.10	99.99997	0.00003	0.3	5.10	99.9841	0.01591	159
5.20	99.99998	0.00002	0.2	5.20	99.9892	0.01078	108
5.30	99.999988	0.000012	0.12	5.30	99.9928	0.00723	72.3
5.40	99.999993	0.000007	0.07	5.40	99.9952	0.00481	48.1
5.50	99.999996	0.000004	0.04	5.50	99.9968	0.00317	31.7
5.60	99.999998	0.000002	0.02	5.60	99.9979	0.00207	20.7
5.70	99.9999988	0.0000012	0.012	5.70	99.9987	0.00133	13.3
5.80	99.9999993	0.0000007	0.007	5.80	99.9991	0.00085	8.5
5.90	99.9999996	0.0000004	0.004	5.90	99.9995	0.00054	5.4
6.00	99.9999998	0.0000002	0.002	6.00	99.9997	0.00034	3.4

Index

D

E

F

G

H

I

J

K

L

M

About the Authors

ERICA L. FARMER, MASTER BLACK BELT

Financial Services Industry

Erica has over 20 years' experience in business process management in the financial services industry. Erica uses her experience to develop others and improve processes through statistics and extensive data analysis. She has been a facilitator and leader of multiple lean and Six Sigma projects. She has experience developing materials and training for Yellow Belt and Green Belt certifications. She has contributed chapters to several quality-related publications. Erica holds a Masters in Organizational Management from the University of Phoenix and a Masters Certificate in Applied Statistics from Penn State University. She is an ASQ CMQ/OE, CQIA, CSSBB, LSS Master Black Belt, Master Certificate in Applied Statistics, PMI Agile Certified Practitioner, and a Project Management Professional (PMP).

GRACE L. DUFFY, PRESIDENT

Management and Performance Systems

Grace has over 45 years' experience in successful business and process management in corporate, government, education, and healthcare. Grace uses her experience as president, CEO, and senior manager to help organizations improve. She has authored 16 texts and many articles on quality, leadership, and organizational performance. She is a frequent speaker and trainer and a member of ISO TC 304. Grace holds an MBA from Georgia State University. She is an ASQ CMQ/OE, CQIA, CSSGB, and CQA. Grace is an LSS Master Black Belt, ASQ fellow and Distinguished Service Medalist. Grace is the 2014 *Quality Magazine* Quality Person of the Year and the 2016 recipient of the Asia-Pacific Quality Organization Milflora M. Gatchalian International Women in Quality Medal.

www.ingramcontent.com/pod-product-compliance
Lightning Source LLC
LaVergne TN
LVHW060037210926
848481LV00013B/3890
* 9 7 8 1 6 3 6 9 4 0 3 2 8 *